Water and Wastewater Treatment

Edited by Murat Eyvaz

Published in London, United Kingdom

IntechOpen

Supporting open minds since 2005

Water and Wastewater Treatment
http://dx.doi.org/10.5772/intechopen.80313
Edited by Murat Eyvaz

Contributors
Jessen George, Ashili Severeni, Shivaraju Harikaranahalli Puttaiah, Félix A. López, José F. Alguacil, Jose Gilberto Torres Torres, Zenaida Guerra Que, Ignacio Cuauhtemoc Lopez, Adrian Cervantes Uribe, Juan Carlos Arevalo Perez, Hermicenda Perez Vidal, Alejandra E. Espinosa De Los Monteros Reyna, Jose Guadalupe Pacheco Sosa, María A. Lunagómez Rocha, Cecilia Sánchez Trinidad, Muhammad Wakil Shahzad, John Murnane, Tony Pembroke, Tom O'Dwyer, Bashir Ghanim, Ronan Courtney, Lisa O'Donoghue, Juan C. Arévalo Pérez, Durvel De La Cruz Romero, Ignacio Cuauhtémoc López, Emmanuel Kweinor Tetteh, Sudesh Rathilal, Maggie Manimagalay Chetty, Edward Kwaku Armah, Dennis Asante-Sackey, Rossitah Selamat

First published in London, United Kingdom, 2019 by IntechOpen
IntechOpen is the global imprint of INTECHOPEN LIMITED, registered in England and Wales, registration number: 11086078, The Shard, 25th floor, 32 London Bridge Street
London, SE19SG – United Kingdom
Printed in Croatia

British Library Cataloguing-in-Publication Data
A catalogue record for this book is available from the British Library

Additional hard and PDF copies can be obtained from orders@intechopen.com

Water and Wastewater Treatment
Edited by Murat Eyvaz
p. cm.
Print ISBN 978-1-78923-929-4
Online ISBN 978-1-78923-930-0
eBook (PDF) ISBN 978-1-78984-688-1

For EU product safety concerns:
IN TECH d.o.o., Prolaz Marije Krucifikse Kozulić 3, 51000 Rijeka, Croatia, info@intechopen.com or visit our website at intechopen.com.

Meet the editor

Dr. Murat Eyvaz is an assistant professor at the Environmental Engineering Department at Gebze Technical University (GTU). He received his BSc degree in Environmental Engineering from Kocaeli University in Turkey in 2004. He completed his MSc and PhD in 2013 at Gebze Institute of Technology (former name of GTU) in Environmental Engineering. He completed his post-doctoral research at the National Research Center on Membrane Technologies in 2015. His research interests are water and wastewater treatment, electrochemical processes, filtration systems/membrane processes, and spectrophotometric and chromatographic analyses. He has coauthored numerous journal articles and conference papers and has taken part in many national projects. He serves as an editor in 30 journals and a reviewer in 100 journals indexed in SCI, SCI-E, and other indexes. He has four patent applications on wastewater treatment systems.

Contents

Preface

The use of water, one of the most valuable and vital resources in the world, should respond to growing needs, and used water should not have negative effects on the environment. Research on the reduction of used water and wastewater quantities, post-use treatment, or reuse/recovery methods is increasing day by day. These studies focus on finding the most appropriate method from both technical and economic perspectives. To reduce the adverse environmental effects of sludge or to reduce the amount of treatment chemicals used, biocides, chelating agents, and fouling cleaners are being developed. Energy and electricity requirements for water and wastewater treatment are being supplied from renewable energy sources instead of fossil fuels. Moreover, energy production from water and wastewater treatment processes is being utilized. The widespread use of new technologies and materials in water and wastewater treatment also reduces operating costs. In addition to the zero liquid discharge approach after treatment, zero solids discharge is now being investigated.

In this book, emerging technologies and materials used in the treatment, reuse, or recovery of various kinds of water and wastewaters are examined. This book consists of eight chapters specifically including desalination and use of renewable energy, nanomaterials, biosorbents, photocatalytic treatment, as well as riverbank filtration and wetlands.

Murat Eyvaz
Assistant Professor,
Department of Environmental Engineering,
Gebze Technical University,
Turkey

Chapter 1

Treatment of Water and Wastewater for Reuse and Energy Generation-Emerging Technologies

Emmanuel Kweinor Tetteh, Sudesh Rathilal, Maggie Chetty, Edward Kwaku Armah and Dennis Asante-Sackey

Abstract

Fresh water quality and supply, particularly for domestic and industrial purposes, are deteriorating with contamination threats on water resources. Multiple technologies in the conventional wastewater treatment (WWT) settings have been adopted to purify water to a desirable quality. However, the design and selection of a suitable cost-effective treatment scheme for a catchment area are essential and have many considerations including land availability, energy, effluent quality and operational simplicity. Three emerging technologies are discussed, including anaerobic digestion, advanced oxidation processes (AOPs) and membrane technology, which holds great promise to provide integrational alternatives for manifold WWT process and distribution systems to mitigate contaminants and meet acceptable limitations. The main applications, basic principles, merits and demerits of the aforementioned technologies are addressed in relation to their current limitations and future research needs in terms of renewable energy. Hence, the advancement in manufacturing industry along with WWT blueprints will enhance the application of these technologies for the sustainable management and conservation of water.

Keywords: anaerobic digestion, advanced oxidation processes, membrane technology, renewable energy

1. Introduction

Wastewater, which is the biggest waste stream from municipalities, petrochemical, pharmaceuticals, food, textile, agricultural, polymer and paper industries and so on contain high contaminants of oil and salts of organic and inorganic compounds [1–5]. This strikes as a major ecological problem with high environmental impacts when discharged into the ecosystem without proper treatment. Furthermore, the industrial revolution associated with demographic growth have increased the demand for freshwater supply, which is depleting the natural fresh water supply sources [3, 5], although wastewater can be treated through various physical, chemical and biological strategies [1–3]. Unfortunately, the current conventional wastewater treatment methods cannot eliminate the contaminants. In addition conventional wastewater treatment can be expensive. Therefore, the quest for clean water and clean environment has resulted in various environmental protection agencies setting stringent discharge limits [3, 4]. Conversely, there are always variations in wastewater qualities which have different impacts on the environment [2–4], where a proper wastewater treatment incorporated with primary, secondary and advanced treatment strategies seems to more viable [1, 3, 4]. The primary treatment involves separating the solids from the liquids via filtration or sedimentation, whereas the secondary treatment removes the dissolved solids and other contaminants through chemical precipitation and biological process [4, 6, 7]. Then UV light or membranes are used for further treatment [1, 2, 5]. After which, the treated water can be profitable to farmers as well as the environment positively in sustainable manner viz. irrigation, and agricultural purposes. In this study, evaluating the streamline flow of innovative wastewater treatment technologies for reuse and subsequent sludge generation as an energy source is being addressed. The biological treatment is presented in section one, followed by membrane technology and lastly the advanced oxidation process. The current limitations and future prospects of each technology are also presented.

2. The biological wastewater treatment process

Municipal solid wastes are attracting more obstructive legislation with respect to landfill disposal of the biodegradable fraction [4, 8]. The treatment process for these organic fractions is biological wastewater treatment. These technologies maximises the recycling and recovery processes of waste components. The biological treatment is regarded as an important and vital aspect of wastewater treatment and is a technique employed for municipal or industrial use for soluble organic components [9]. Among all, the most widely employed method for sludge treatment is anaerobic digestion [9, 10].

2.1 Anaerobic digestion (AD) process

In this process, a large fraction of the organic matter (cells) is broken down into carbon dioxide (CO_2) and methane (CH_4), and this is accomplished in the absence of oxygen. About half of the amount is then converted into gases, while the remainder is dried and becomes a residual soil-like material. Kougias and Angelidaki [11] reported that the end products of organic assimilation in anaerobic treatment of waste are CH_4 and CO_2 as depicted in **Figure 1**. The AD technology has encountered significant recognition in the last few decades with the applications of separately configured high rate treatment processes for industrial wastewater streams. In the wastewater treatment settings, the AD has been employed in several instances throughout the world for bioremediation and biogas production [8, 12, 13]. Biogas, a well-known and common renewable source of energy, is produced via the AD process, consisting largely of CH_4 and CO_2. As an alternative source of energy, the AD process produces biogas that can be chiefly used as fuel in combined heat and power gas engines [11, 12, 14]. There has also been a rapid adoption of anaerobic co-digestion, where two or more different feed stocks are digested together in anaerobic biodigesters with the core aim of improving the biogas yield [8, 11–13, 15–17]. Other advantages ensured in the anaerobic systems include lower energy requirements, a safer and more convenient way of converting "waste" to useful products associated with urbanisation, being a predictive tool for the fulfilment of the UN Sustainability Goal to meet Global standards, having excellent nutrient recovery and high organic removal efficiencies. Drawbacks include longer hydraulic retention times (2–4 months) and high alkalinity requirements [11, 18, 19]. The aerobic system presents merits such as high organic removal efficiencies, excellent effluent quality

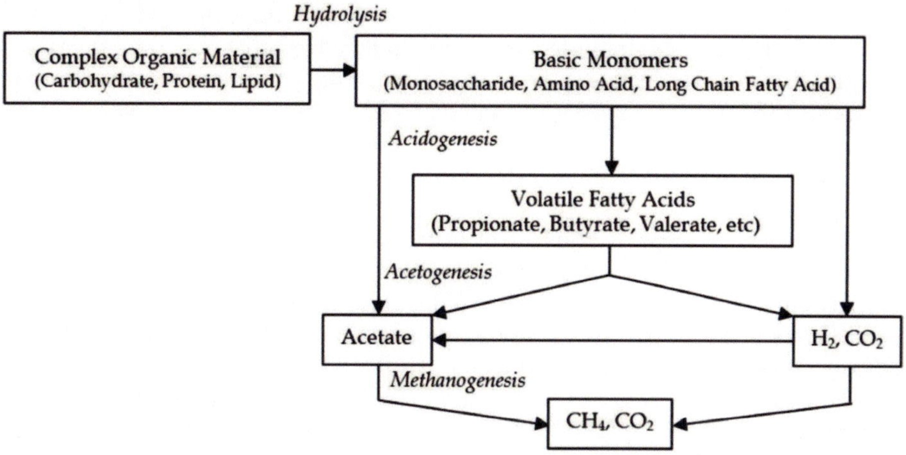

Figure 1.
Modified stages of the anaerobic digestion process, adapted from [11, 12].

and shorter start-up times (2–4 weeks). Demerits include longer hydraulic retention times, pretreatment requirements for delignification of lignocellulosic biomass, odour built-up in bioreactors, costs associated with CO_2 upgrading, no nutrient recovery and high energy requirements [8, 12, 17]. Research conducted by Kainthola et al. [19] details the major differences between the anaerobic and aerobic systems of wastewater treatments, as depicted in **Figure 2**, demonstrating the mechanism with species required and products formed. The anaerobic/aerobic systems have also been employed largely at both municipal and industrial levels as a method for wastewater treatment for many years. It presents advantages such as a lower consumption of energy, low chemical consumption, low sludge production, its enormous potential for the recovery of resources, simplicity of the operation and the requirement of less equipment. Some advantages of the biological treatment method over other treatment techniques such as thermal and chemical oxidations are capital investments required and costs in operation of the processes [8, 11, 12, 14].

2.1.1 Operating parameters of AD process

Some operating parameters which are usually monitored and optimised to maximise the performance and operation of AD include organic loading rate, pH, hydraulic

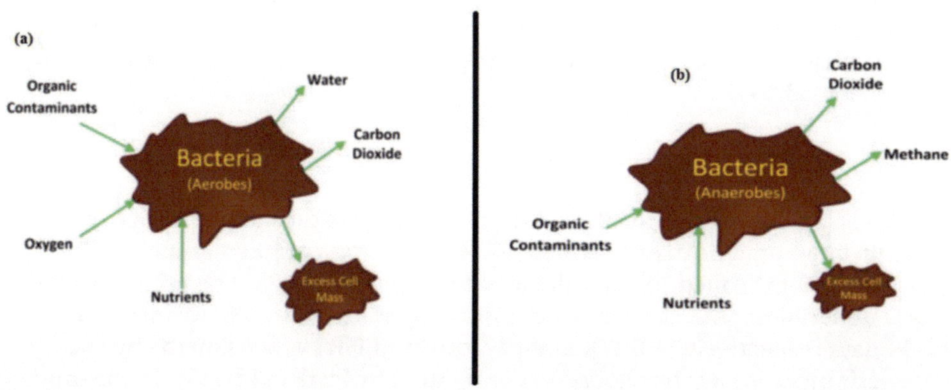

Figure 2.
Schematic diagrams of (a) aerobic treatment principle and (b) anaerobic treatment principle, adapted from [11, 12].

retention time, temperature, carbon to nitrogen ratio and many more [15, 20]. As a result, any sharp variation in these parameters could adversely affect the substrate concentration in the biodigesters. Some of the operating parameters are discussed in Sections 2.2.1–2.2.4.

2.1.1.1 Organic loading rate (OLR)

This is generally expressed in terms of the amount of chemical oxygen demand (COD) or volatile solids (VS) of digester volume in a day and denoted as $KgCOD/m^3$ d or $KgVS/dm^3$ d. Most favourable COD removal of the canning industry effluent was found to be between 89 and 93% at OLRs of 9.8 and 10.95 $kgCOD/m^3$ d at an HRT of 10 h at a pH of 5.5 [15, 18, 19]. This prediction becomes viable during the selection of the reactor-type and other process parameters such as pH control. OLR has been found to increase with decreasing biodegradation of the volatile solid and the subsequent bioenergy produced. The performances of bioreactors decrease when the OLRs increase with energy production [10, 15, 17].

Furthermore, the pH range suitable for AD is reported to be within the range of 6.8–7.2 [13]. This is achieved by charging the AD at an optimum OLR to obtain a higher yield of biogas. There is usually a variation in pH during AD especially during acidogenesis where volatile fatty acids such as propionate, butyrate and acetates are produced [19]. The presence of phosphates (PO_4^-) in most wastewater treatment facilities renders the pH adjustment with calcium hydroxide possible to a pH of about 7.2, even at high concentrations. The growth of microorganisms in AD is largely dependent on the pH of the substrates undergoing the overall biodegradation [8]. In the treatment of wastewaters, the observed pH range of 6.0–7.1 was been reported in a study where a mixed batch reactor produced larger quantities of biogas at an average value of 0.405 m^3/d [11].

2.1.1.2 Temperature

Temperature conditions during the AD process for bioenergy production includes psychrophilic (<30°C), mesophilic (30–40°C) and thermophilic (50–60°C) [8, 12]. The anaerobes are found to be more active under both mesophilic and thermophilic temperatures as compared to psychrophilic temperatures. Comparatively, thermophilic temperatures are considered suitable for the enhancement of biomethanation by accelerating the hydrolysis of the polymeric feedstock and other metabolic pathways [12, 13]. However, several studies have shown that thermophilic digesters suffer from poor process stability due to volatile fatty acid accumulation during the acidogenesis process, most especially propionate [13, 17].

2.1.1.3 Hydraulic retention time (HRT)

This is the measure of the time required to achieve the complete biodegradation of an organic matter associated with process parameters such as the temperature of the medium and the waste composition [12]. The HRTs observed in AD under mesophilic and thermophilic temperatures are 15–30 days and 12–14 days respectively [13, 17]. Temperature and HRT effects on the methanogenesis process have been observed in a study by Shah et al. [13]. In the same study, the working temperature was adjusted from 30 to 55°C following an HRT of 8–12 days.

2.1.2 Applications of the anaerobic digestion process

The full-scale application of the AD technique in the treatment of industrial wastewater depends on the hydrodynamic configuration of the AD reactor. There have been different types of AD reactors applicable in the wastewater treatment settings, which includes continuous stirred tank reactor (CSTR), the anaerobic sequencing batch reactor (ASBR), upflow hybrid anaerobic sludge-filter bed (UASFB), upflow anaerobic sludge blanket reactor (UASBR), expanded granular sludge bed (EGSB), anaerobic baffled reactor (ABR), anaerobic fixed-bed reactors (AFBR) and integrated bio-membrane reactors [8, 13, 15–17]. For instance, **Figure 3** depicts schematic cross section view of the upflow hybrid anaerobic sludge-filter bed and the upflow anaerobic sludge blanket reactor. The integrated anaerobic-aerobic bioreactors have been most preferred in the past few decades due to its ability to meet stringent constraints in terms of mitigating odorant compound release and minimising sludge production.

2.1.2.1 Limitations of adapting AD process in large scale

Research has shown that among the various reactors used in the performances for the treatment of wastewaters, the UASBR configuration is the most widely used with a high-rate anaerobic reactor for the treatment of high-strength wastewater [8, 12]. Several modifications have been carried out in the design of bioreactors to enhance both the consistency and the efficiency of the reactors. The AD process does encounter failures causing serious environmental hazards [8]. In addition, some of the aforementioned operating parameters as previously discussed (Section 2.2) can affect the performances of the microbes responsible for the biodegradation of organic matter in the wastewater settings. Further drawbacks observed in AD large scale operation include microbial shift, process instability, low yield of biogas production and poor water quality [8, 10, 12, 14, 19]. For instance, monodigestion of energy crops still struggle to meet the reduction targets concerned with the drawbacks in AD compared to anaerobic co-digestion (AcoD) such as a mixture of slurry and energy crops [5, 20]. In response, pretreatment techniques for cellulose enhancement and the use of energy crops as feed stocks have been found to increase the efficacy of biogas production via AD [12]. Some of the improvement techniques which have gained attention in terms of research for the betterment of AD process design and the optimisation includes

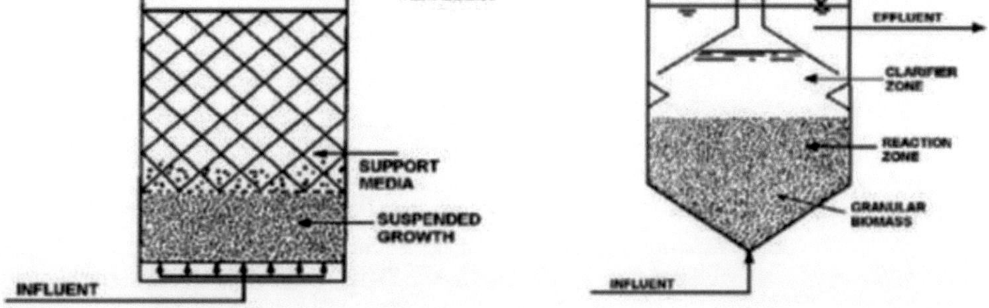

Figure 3.
(a) Upflow-hybrid anaerobic sludge-filter bed and (b) upflow-anaerobic sludge blanket reactor [8].

evaluating the AD process kinetics and dynamics, nitrification-denitrification, recycling of the centrate back to the AD reactor, wastewater characterisation, optimisation of operational and environmental parameters, and microbial community shift.

2.1.3 Future prospects of AD application

According to the United Nations Sustainability Development Goal of 2030, the use of renewable energy is expected to reach 100% by the year 2050. The seventh goal focuses on the production of affordable and clean energy globally which is environmentally friendly [21, 22]. Renewable energy has gained attention to cater for the ever-increasing use and over-reliance of non-renewable forms of energy. This arises because of the emission of greenhouse gases compelling researchers in the past decades to search for an alternative means of sustainable energy production [17, 23–25]. The reserve for energy has become necessary for global concern in maintaining a sustainable way in lieu of the resources available especially at WWTPs. Aside the protection of the environment, wastewater treatment plants (WWTPs) also serve as a source of generating renewable forms of energy such as biogas. In a WWTP, the dumping of sewage sludge produced as a by-product is a problem of growing significance representing up to 50% of the entire operating costs of all WWTPs [8, 12, 14]. Also, wastewaters with a high content of nitrogen can be treated with the nitrification and denitrification technique form of AD generally known as Anammox [26]. Constructive government policies have shown Germany as being the dominant global biogas energy generation country globally for the future [23, 25]. Latest reports predict that biogas production could increase from 18,244 Gigawatt hours (GWh) in 2012 to 28,265 GWh in 2025, indicating a compound annual growth rate (CAGR) of 3.4%. In 2011, Germany contributed the largest share of the world's cumulative installed capacity as the country accounted for approximately a quarter of the global biogas biogas [23, 25, 27].

3. Membrane technology

Membranes, as a thin layer barrier for size differential separation, are usually integrated with chemical and biological treatment or standalone systems in secondary treatment of wastewater settings [28–30]. In a typical membrane mechanism, there is usually a driving force, such as a semi-permeable barrier which controls the rate of movement of components by fractional permeation and rejection through pores of different sizes as depicted in **Figure 4** [32]. The permeation and selective rejection is a function of the membrane pore size and chemical affinity, which helps

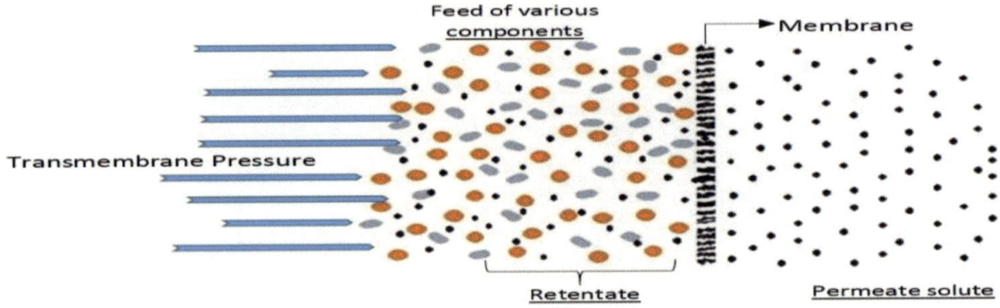

Figure 4.
Membrane selective permeation for various solutes adapted from [31].

to have a product stream devoid of target components [33]. Due to the relatively low energy requirement and wastewater treatability efficiency, membrane technology has tremendously improved by the development of new materials and configurations for industrial applications. Some of these applications include microbial fuel cells, removal of organic and inorganic components, disinfection, pathogen removal and desalination [30, 33, 34].

3.1 Types of membrane technologies

Generally, the major driving force for selective filtration is a potential gradient of variables such as hydrostatic pressure, electrical voltage, temperature, concentration or a combination of these driving forces [29, 32]. These variables including nature (natural and synthetic) and structure (porous or non-porous and heterogeneous or homogenous) have been used in the classification of membranes [28, 31, 32, 34]. However, most commercially available and industrially used membranes are pressure-driven and energy driven (electrodialysis and electrodialysis reversal) membranes [35]. Pressure driven types are namely microfiltration (MF), ultrafiltration (UF), reverse osmosis (RO) and nanofiltration (NF). These are also classified by their pore sizes or molecular weight cut-off (MWCO). MWCO is expressed in Daltons (1 Da = 1 g mol^{-1}) and is the minimum or smallest component that can be retained with at least 90% efficiency [28, 30, 32, 35]. It should be noted that as the pore size of these membranes decrease, the driving force for the operation increases. For instance, the MF and UF are referred as low pressure driven processes while RO and NF are high pressure driven processes [28, 29, 34].

3.2 Applications of membrane technology

3.2.1 Microfiltration (MF) application

Microfiltration utilises a sieving mechanism to retain macromolecules or particles more than 0.1 μm, specifically in the range of 0.1–10 μm [30]. Unlike UF, RO and NF, the transmembrane pressure (TMP) for both sides of the membrane as a result of the small particle retention is puny, hence requiring a relatively small TMP lower than 2 bar but varies from 0.1 to 2 bar [28, 30]. As indicated, the larger pore sizes of MF membranes limit removal to suspended solids, bacteria, some viruses (up to 2-log), protozoan cyst, turbidity and on a lesser extent, organic colloids within the region [28, 29, 32].

3.2.2 Ultrafiltration (UF) application

The role of UF is increasing as a pretreatment for desalination and membrane bioreactors. Ultrafiltration (UF) like MF utilises physical sieving as a separation mechanism. The pore size, MWCO and pressure for a UF membrane ranges from 0.05 μm to 1 nm, 1–500 kDa and operating pressure of 1–7 bar [30, 33]. As such, UF with a definite MWCO are impermeable to compounds with molecular weights exceeding the MWCO and have shown a 3–6 log removal of chlorine resistant protozoan cysts, active *Giardia lamblia*, colloids, viruses and coliform bacteria. The use of MF and UF as pretreatment to RO has gradually emerged as an industry standard. Both are often used as pretreatment for NF and RO processes to reduce membrane fouling and is also applied as a post treatment to chemical precipitation for organic chemical removal and pH adjustment, phosphorus, hardness and metals [29–31, 33, 34]. Fouling is highly eminent in UF due to the high molecular weight of fractions retained in relations with the small osmotic pressure differentials and liquid phase diffusivity. However, this does not negatively influence the demand for UF's as any design, configuration and application will be fouled [28, 30, 36]. The configuration for application is influenced by the mechanical stability, hydrodynamic requirement and economic limitations.

3.2.3 Nanofiltration (NF) application

Nano filtration is a pressure related process where the mechanism of separation is based on molecular size for the removal of dissolved micro pollutants and multivalent ions. The NF is a complex process characterised by solvent diffusion, transport and electrostatic repulsion effects at the membrane surface and within the nanopores [29, 30]. The difference between the pore diameter and particle size forms the basis of the sieve effect. Based on the membrane pore size, NF is often referred to as 'loose RO' with separation taking place at the lower end of UF and upper end of RO as it covers a MWCO of 100–500 Da, a pore size of 0.5–5 nm and operates at relatively low pressure of 5–35 bar [28, 30, 34]. The NF is usually deployed in the removal of polyvalent cation, reduction in colour, tannins, turbidity and disinfection by-product precursors such as organic matter as their potency lies in the high rejection of divalent ions (98%), permeation of monovalent ions (20–80% rejection) and high flux. Nano filtration is often used as a post treatment or polishing step in conventional treatment processes. Although it is not advisable to be used in desalination processes, it is used to reduce the salt content of slightly saline water. Recent applications have used NF as a pretreatment to RO reducing the operating pressure in RO providing savings in operational and maintenance costs [31, 32, 35]. Second stage fouling is usually reduced in NF systems through ozone pretreatment and non-thermal crystallisation while cleaning is done using suitable alternatives that also exist for MF and UF [30, 33].

3.2.4. Reverse and forward osmosis

Reverse osmosis, often referred to as tight membrane has been widely used in brackish water and wastewater treatment with its effectiveness in desalination against conventional thermal Multi stage flash. High external pressure of 15–150 bar which is a function of the hypertonic feed and greater than the osmotic pressure is applied to retain dissolved solute and solvent permeation at a MWCO around 100 Da through diffusion mechanism [37–39]. Using concentration gradient as the driving force, separation and concentration in forward osmosis (FO) occurs as the concentrated solution (e.g., salts such as NaCl) draws water from a less concentrated feed solution. Characteristic advantages include low energy consumption, simple configuration and operation, low membrane fouling tendency and high rejection of a wide range of contaminants. The use of FO operates at ambient conditions, hence irreversible fouling is low [37, 39]. However, FO technology is faced with the lack of recyclable and economical drawing solutions, internal and external concentration polarisation and a difficulty of developing effective large scale plants [37]. To achieve desired process flow and optimum configuration, ROs are arranged in stages and passes. The sequence of the stages has the concentrate stream of the first stage as the feed inlet of the second stage. Permeate streams from both streams are summed into one discharge channel. However, passes involve either a one path recovery of permeate or the rechannelling of permeates from the first RO into the second RO to improve quality [40, 41] as summarised in **Table 1**.

Membrane	Microfiltration	Ultrafiltration	Nanofiltration	Reverse osmosis
Structure	Asymmetrical (thickness 10–150 μm)	Symmetric (thickness = 10–200 μm) or asymmetric (thickness = 0.1–0.5 μm and supported by sublayer of 50–150 μm)	Asymmetric 150 μm	Asymmetric 150 μm
Material	Ceramic, PP, PSO, PVDF	Ceramic, PTFE, PSO, CA, PVDF, thin film	PA,CA, PES SPSF, PI, PVA, CS, organic/ inorganic hybrids: thin film	Zeolite, PVDF, PES, PBI, TFC
Transport Laws	Darcy' Law	Darcy	Fick's	Fick's
Module	Dead end: flat sheet or spiral wound Cross flow: capillary, hollow fibre or tubular	Capillary, spiral wound, others: tubular, plate and frame, rotary modules, vibrating modules and dean vortices	Tubular, spiral wound, plate and frame	Tubular, spiral wound, plate and frame
Configuration	Dead end, cross flow, transverse flow	Dead end, cross flow, transverse	Shell-side feed, bore-side feed, cross flow	Cross flow in various stages and passes

PBI, polybenzimidaole; CA, cellulose acetate; PSO, polysulfone; PP, polypropylene; PVDF, polyvinylidene fluoride; PES, polyethersulfone; TFC, thin film composite.

Table 1.
Summary of pressure-driven membranes [39–41].

3.2.5 Membrane bioreactors

Low-pressure driven MF and UF used for critical solid-liquid separation has been integrated with biological treatment into a hybrid activated sludge process termed as membrane bioreactors (MBR) for wastewater treatment. Unlike conventional wastewater treatment processes with treatment limitations, MBRs have shown wide range treatment efficiency in the removal of organic and inorganic emerging micro pollutants including pesticides, antibiotics, analgesics, anti-epileptic, biodegradable organic compounds, microplastics, industrial chemicals and nutrients [30, 42, 43]. The MBRs can be configured into gravity or submerged vacuum driven systems by using hollow fibre or flat sheet modules, while pressure driven membranes on the other hand are mostly configured with external pipe cartridge systems. The MBR is usually combined with conventional systems including thermophilic or mesophilic bioprocesses, AOPs, powdered activation carbon to enhance improve the water quality and treatability efficiency [39, 40, 42, 44]. Some of the advantages of MBRs include small foot-print requirement, simple transport configuration and the ability to handle high biomass concentrations [45].

3.2.6 Ion exchange membrane

Ion exchange membranes are classified as anion exchange membrane (AEM) if the polymer matrix is embedded with fixed positive charge groups and vice

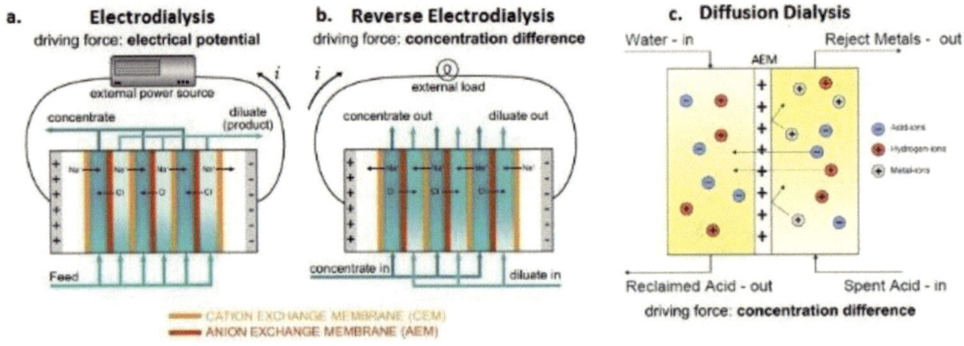

Figure 5.
Configurations of (a) electrodialysis, (b) reverse electrodialysis and (c) diffusion dialysis [35].

versa for cation exchange membranes (CEM) [35]. This involves the permeation of anions/cations and rejection of cations/anions in the effluent. Electrodialysis (ED), reverse electrodialysis (RED), diffusion dialysis (DD) and Donnan membrane process (DMP) are examples of these, which usually involves the exchange of ions between the solutions across the membrane as shown in **Figure 5**. The application of these processes is usually based on the type of effluent and is reported to be non-fouled and energy resourceful as mechanism of separation is by potential gradient. The DD is applied to reclaim free mineral acids and alkalis while DMP is used in recovery of toxic and valuable metals from various feeds. In both DD and DMP, a Donnan equilibrium is established hence their difference lies the in application areas. Unlike ED and RED, DD and DMP are being applied on bench scale, have osmotic limitations and therefore kinetic studies are been considered for various effluents. However, ED and RED constructions requires compatibility of feed stream and stack materials, electrical safety consideration and larger footprint to produce equivalent water quantity and quality [30, 32, 33, 36].

4. Advanced oxidation process

Basically, there are two stages which are usually employed in wastewater treatment settings via a pre-treatment step involving mechanical and physicochemical systems to reduce the heterogeneous components of the effluents followed by an advanced treatment process. The physiochemical process enhances the efficiency of the advanced treatment by agglomerating the containments into a larger size for easy filtration or removal [46, 47]. However, degradation of emerging recalcitrant components with membrane and bioremediation in advanced treatment processes attests to be complex. So, in addressing this problem, advanced oxidation process (AOP) has gained much attention due to its potential to degrade a wide range of organic micro-pollutants [47, 48]. This process involves the generating of potent reactive hydroxyl radicals (E_0 = 2.8 eV) with photon energy and without further additional chemical treatment. Examples are chemical oxidation (O_3, Fenton reagents), photochemical oxidation (Ultraviolet-UV/O_3, UV/H_2O_2), heterogeneous photocatalysis (UV/TiO_2), electrolysis and sonolysis [46, 49–52] as shown **Figure 6**. These technologies use UV-A with long wavelengths of 315–400 nm, and UV-C with short wavelength radiation of 100–280 nm for degradation of most environmental contaminants. Generally, UV/O_3 and UV/H_2O_2 processes consume large amounts of oxidant, which makes them uneconomical to operate [43]. On the other hand,

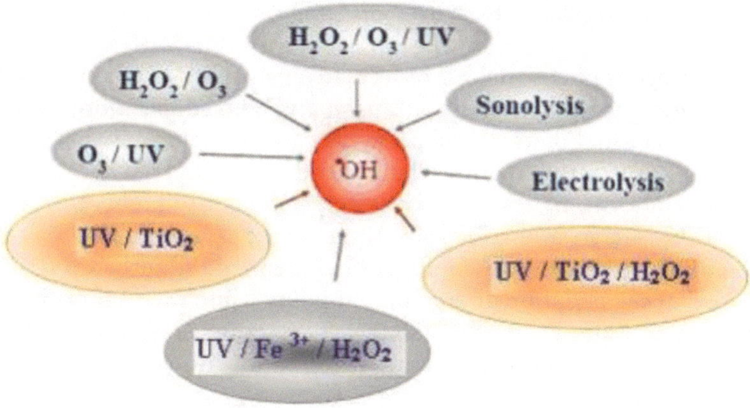

Figure 6.
Types of advance oxidation process (AOPs) adapted from [46].

the hazards associated with ozone being an unstable gas limits its application and is usually coupled with an ozone-water contacting device to convert the ozone into its liquid phase thus increasing the cost of production. However, the considerable operational conditions of ambient temperature and pressure and the use of a low-cost and chemical stable catalyst (TiO_2) are predominantly attractive for complete mineralisation of contaminants and by-products. This makes heterogeneous photocatalysis techniques to be advantageous over other AOP's. Other advantages include no sludge production, quick reaction rate, low cost and operating well at ambient temperature and pressure conditions [46, 49].

4.1 Photocatalysis

There are several semiconductors favourable for water treatment such as TiO_2, ZnO, Fe_2O_3 CdS and ZnS, which are active within a band gap energy of 2.3–3.2 eV and wave length of 413–539 nm [53, 54]. TiO_2 has been the most widely used photocatalyst and exists in three major crystalline polymorphs such as anatase, rutile and brookite (**Figure 7**). There have been intensive studies to determine the crystalline structure, surface area, density of surface hydroxyl groups and adsorption/desorption characteristics of TiO_2 [46, 49]. As mentioned, the surface hydroxyl group concentration has been documented to play a vital role in the photocatalytic degradation process, such that an increase in concentration of the hydroxyl groups on the catalyst surface might have a positive effect on the reaction rate [46, 49, 51]. Thus, the hydroxyl radicals produced (OH^-) are a very combative species and are therefore able to oxidise a wide range of organic pollutants in a swift and effectual means.

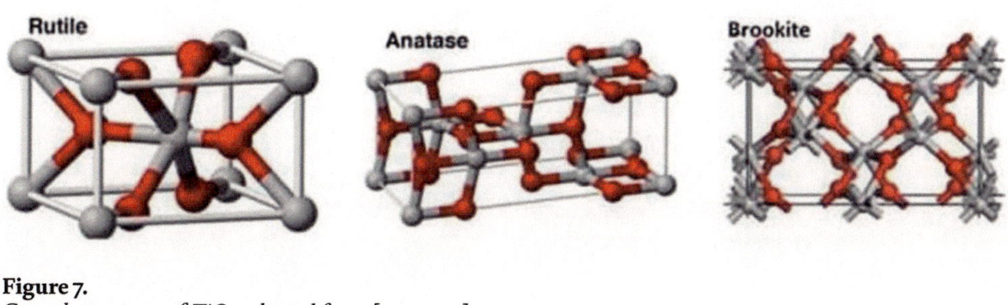

Figure 7.
Crystal structures of TiO_2 adapted from [51, 52, 55].

4.1.1 Principles of heterogeneous photocatalysis

The heterogeneous photocatalysis process involves series of reactions such as oxidation, dehydrogenation, hydrogen transfer and metal deposition, among others [55]. Besides, the normal thermal and chemical catalytic actions that ease any distinctive chemical reaction require simultaneous activation of the molecules. As such, in photocatalysis, a light source with energy equal to or higher than the band-gap energy of the catalyst is employed to stimulate the catalyst to be active during the reaction bringing about the reduction or oxidation of the adsorbed species on the surface, for instance, oxidation of organic compounds into their subordinates until carbon dioxide and water are formed [56, 57]. This results in charge separation as the electron (e^-) moves from the valence band to the conduction band of the semiconductor catalyst, resulting in a hole (h^+) in the valence band as shown in **Figure 8**.

4.1.1.1 Selection of catalysts

Some of the parameters to consider in selecting a photocatalyst to enhance the degradation capacity include the nature and intensity of light source, the amount of photons to activate the catalyst, the reaction medium and the water chemistry to generate the hydroxyl radicals, and the nature and concentration of the pollutants [46, 51, 55]. However, the relationship between the energy levels of conduction and valence bands with respect to the energy for reduction and oxidation is one of the cardinal points to consider in selecting a photocatalyst [46, 47]. The chemical capacity of the photo-generated electrons and holes has high influence on the conduction band energy level of the photocatalyst. Thus the photo-generated electron of a catalyst should be able to

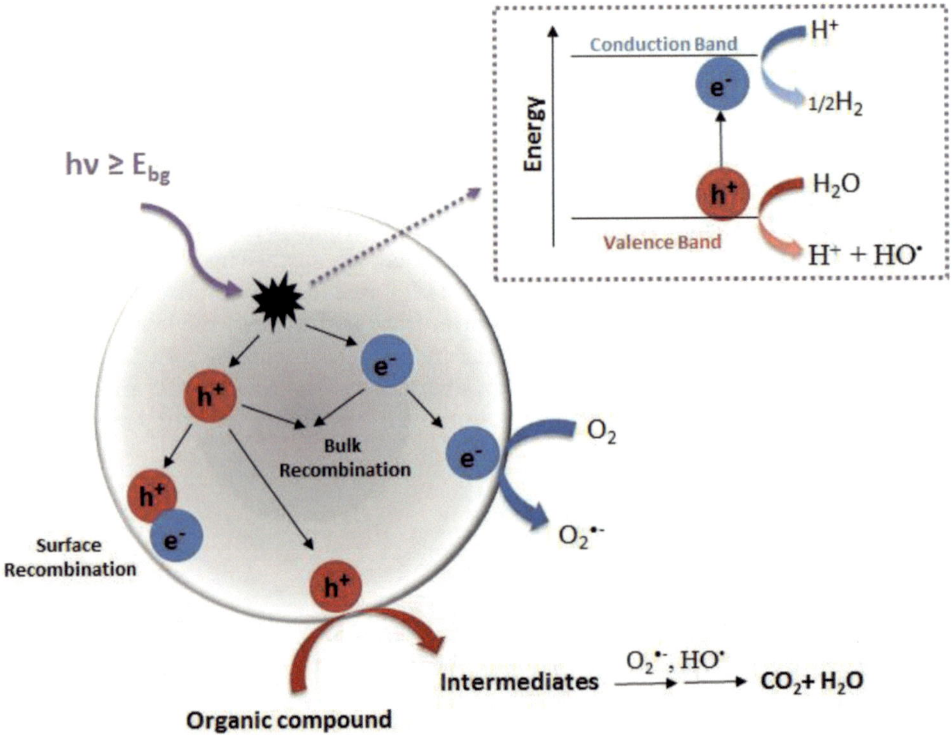

Figure 8.
Schematic illustration of the photocatalytic mechanism in the presence of a water contaminant adapted from [46, 49, 52].

have essential energy to ease the mineralisation of the micro pollutants. The conduction band energy level of the photocatalyst in terms of mineralisation has to be more negative, while the valence band has to be more positive with respect to the energy for oxidation of water. This makes semiconductors with large band gaps to be considered as suitable for photocatalytic activities [46, 47, 50]. Thus, semiconductor materials are able to provide adequate negative and positive redox potentials in conduction bands and valence bands, respectively. In addition, the wider the band-gap of semiconductor material the higher the energy input to enhance its efficiency. Furthermore, the poor photo-corrosion stability of many semiconductors limits significantly the number of potential photo-catalytic materials for mineralisation in wastewater settings [55, 57].

4.1.1.2 Modification of catalyst

Photocatalytic activity of a distinctive semiconductor, such as TiO_2, can attract light energy equivalent to or higher than the band gap energy, resulting in generating electrons and holes. This electron-hole pair then migrates to the TiO_2 surface and amalgamates with the adsorbed reactants to hasten the reduction and oxidation process. The lack of such an energy transfer leads to recombination of the pairs, which then competes with the desirable redox process with a high loss of energy. Therefore, to increase photocatalysis efficiency, suitable modification of the semiconductor band gap is essential to attract the recombination charge carriers which might hinder the photocatalytic efficiency. Some of these processes include (i) doping with metal cations and anions, (ii) coupling with other semiconducting oxides, (iii) sensitization with light harvesting compounds/dye molecules and (iv) plasmon resonance induced by specific metals [49, 51].

4.1.2 Parameters affecting photocatalysis

As with most processing techniques, several parameters affect the performance of the photocatalytic degradation process. Photocatalytic activities are being influenced by experimental conditions such as the amount of catalyst, light intensity, lighting area, reactor volume, pH type of reactor, temperature and pressure. Likewise, the inherent structural and properties of semiconductor photocatalyst influences its performance. This includes the particle size, phase composition, surface area, surface hydroxyls, lattice defects and the type of dopants (metals and non-metals) [46–48, 52]. The amount of light radiation absorbed is the driving force for semiconductor with threshold wavelength to provide the appropriate photon energy to overcome the band gap between the valence and conduction bands. This threshold wavelength is very important to promote electrons to be in an excited state corresponding to the band gap energy as depicted in **Table 2**.

4.1.3 Applications of photocatalysis

With the current greenhouse emission problems and energy deprivation situation, photocatalysis emerges as one of the alternative ways of providing feasible solutions to the global front in relation to energy and the environment. For instance (**Figure 9**), in the phenomenon of photosynthesis, green plants trap the solar energy from the sunlight and with a series of enzyme catalysed redox processes, converts the CO_2 into water and carbohydrates by releasing oxygen into the atmosphere where most living organisms depend on to survive. In this scenario, photocatalysis which works with the same principle has gained incredible status and hence can be explored for divert applications in seeking sustainable energy, social economic growth and environmental impact.

Semiconductor	Band gap energy (eV)	Wavelength (nm)
TiO_2 (rutile)	3	413
TiO_2 (anatase)	3.2	388
ZnO	3.2	388
ZnS	3.6	335
CdS	2.4	516
Fe_2O_3	2.3	539
Adapted from [47].		

Table 2.
Semiconductors and their band gaps at specific wavelengths.

4.1.3.1 Photocatalytic degradation in wastewater

The importance of photocatalytic degradation technology in pre- and post-treatment of water and wastewater using sunlight on a large scale is at the verge of development, with few of them like self-cleaning, anti-fogging and anti-bacterial applications currently being practiced. For instance, two slurry reactors coupled with TiO_2 were built in New Mexico (USA) and Almeria (Spain); however, the environmental conditions, low photonic yield and efficiency under the visible light make it challenging to be commercialised [27, 60–63]. There are numerous studies ongoing seeking to address some of these limitations in order to improve the photo-catalyst, reactor design and light efficiency [55, 59]. Furthermore, the kinetics study to understand the mechanistic pathways of mineralisation of recalcitrant organic and micropollutants has also received attention.

4.1.3.2 Photocatalytic energy production

In the early 1900s, the photocatalytic techniques commonly referred to as "artificial photo-synthesis" was employed in reducing CO_2 into useful hydrocarbons (**Figure 9**). This has now attained incredible status on the global level. Currently researchers, scientists and engineers are exploring the mechanism to control the current atmospheric carbon dioxide levels (green-house gas effect) by altering it into to fuels and useful chemicals, and the role of CO_2 as a source of energy. However, the

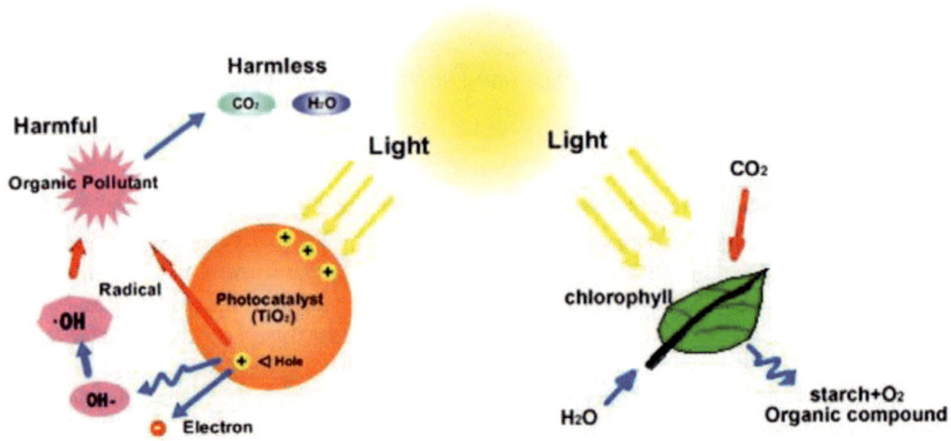

Figure 9.
Schematic comparison of photo-catalysis and photo-synthesis [58, 59].

use of solar energy via chemical storage can be attained by photocatalytic or photo-electrochemical initiation of light-sensitive catalytic surfaces. Due to the simplicity of the photocatalytic process, it can be employed in converting solar energy into other useful forms of energy like hydrogen via splitting of water and hydrocarbons (methane, methanol, etc.), commonly known as solar fuels [52, 55, 59].

4.1.4 Future prospects for photocatalysis

There have been tremendous efforts dedicated to improving photocatalytic degradation efficiency, where some role of the aforementioned parameters on the AOPs performance have been studied with different types of wastewaters from the textile, oil refinery, slaughterhouses [46, 47, 59]. However, in relation to relative operational and energy utilisation costs, as well as the formation of unknown toxic intermediates from the parent compounds (semiconducts) and the pollutants remain unsolved. Although, this method is prone to shifting of hydroxyl radicals by non-target substances, but some are not suitable for certain varieties of toxic compounds that can counterattack the hydroxyl radicals rendering an increase in ecotoxicity when discharged into the environment. In response, there have been attempts to improve the semiconductor material performance by discovering the reaction mechanisms to develop commercial scale technology, but there are still some challenges which include mass transfer limitations, catalyst deactivation, generation of intermediate products and by-products, and the multi-complex optimisation of the materials and the reactor configuration [49, 55, 58]. Furthermore, due to the intrinsic nature of photocatalysis, it is very difficult to model photocatalytic reactors with a uniform light along the whole reactor volume [58, 59]. Therefore, a feasibility study on the irradiance distribution inside the reactor is required, especially with heterogeneous media such as TiO_2 in suspension.

5. Conclusion

The availability of a cost effective technology to ensure the economic viability of wastewater settings for domestic and industrial purposes is still limited. This chapter presented the insights in applicability of some of the cost-effective technologies in addressing the global water, energy and environmental concerns. The fate of biological systems (AD process), and some of its operating parameters essential for wastewater treatment and bioremediation to energy (biogas) were also discussed. Furthermore, limitations, applications and different configuration types of AD reactors, as well as the forthcoming of the AD process as the alternative technology for bioenergy production using wastewater as a source was discussed. The emerging membrane technologies such as Donnan membrane process are also spotted as the foreseen green and energy saving technologies for industrial and environmental applications. The future challenge for using AOPs in wastewater treatment could be addressed by developing a cost effective photocatalyst to enhance the wastewater treatment and coupled in reduction of carbon dioxide as a renewable energy source to fuels. Therefore, adapting strategies for processes integration and commercialisation of the aforementioned technologies will enhance sustainable social economy growth and development.

Acknowledgements

The authors wish to thank Durban University of Technology, National Research Foundation and Water Research Commission, South Africa, for their support.

Author details

Emmanuel Kweinor Tetteh*, Sudesh Rathilal, Maggie Chetty,
Edward Kwaku Armah and Dennis Asante-Sackey
Department of Chemical Engineering, Faculty of Engineering and the Built Environment, Durban University of Technology, Durban, South Africa

*Address all correspondence to: ektetteh34@gmail.com

References

[1] De Andrade JR, Oliveira MF, Da Silva MGC, Vieira MGA. Adsorption of pharmaceuticals from water and wastewater using nonconventional low-cost materials: A review. Industrial and Engineering Chemistry Research. 2018;**57**(9):3103-3127

[2] Freyria F, Geobaldo F, Bonelli B. Nanomaterials for the abatement of pharmaceuticals and personal care products from wastewater. Applied Sciences. 2018;**8**(2):170

[3] Carlos FS, Schaffer N, Andreazza R, Morris LA, Tedesco MJ, Boechat CL, et al. Treated industrial wastewater effects on chemical constitution maize biomass, physicochemical soil properties, and economic balance. Communications in Soil Science and Plant Analysis. 2018;**49**(3):319-333

[4] Rincón Llorente B, De la Lama-Calvente D, Fernández-Rodríguez MJ, Borja-Padilla R. Table olive wastewater: Problem, treatments and future strategy. A review. Frontiers in Microbiology. 2018;**9**:1641

[5] Abbas A, Valek L, Schneider I, Bollmann A, Knopp G, Seitz W, et al. Ecotoxicological impacts of surface water and wastewater from conventional and advanced treatment technologies on brood size, larval length, and cytochrome P450 (35A3) expression in *Caenorhabditis elegans*. Environmental Science and Pollution Research. 2018;**25**(14):13868-13880

[6] Tetteh EK, Rathilal S. Effects of a polymeric organic coagulant for industrial mineral oil wastewater treatment using response surface methodology (Rsm). Water SA. 2018;**44**(2):155-161

[7] Dohare D, Meshram R. Biological treatment of edible oil refinery wastewater activated sludge process and sequencing batch reactors—A review. International Journal of Engineering Sciences & Research Technology. 2014;**3**(12):251-260

[8] Chan YJ, Chong MF, Law CL, Hassell DG. A review on anaerobic-aerobic treatment of industrial and municipal wastewater. Chemical Engineering Journal. 2009;**155**(1-2):1-8

[9] Patil JH, AntonyRaj MA, Shankar BB, Shetty MK, Kumar BP. Anaerobic co-digestion of water hyacinth and sheep waste. Energy Procedia. 2014;**52**:572-578

[10] Tetteh E, Amano KOA, Asante-Sackey D, Armah EKA. Response surface optimisation of biogas potential in co-digestion of miscanthus fuscus and cow dung. International Journal of Technology. 2018;**9**(5):944-954

[11] Kougias PG, Angelidaki I. Biogas and its opportunities—A review. Frontiers of Environmental Science & Engineering. 2018;**12**:1-2

[12] Appels L, Baeyens J, Degrève J, Dewil R. Principles and potential of the anaerobic digestion of waste-activated sludge. Progress in Energy and Combustion Science. 2008;**34**(6):755-781

[13] Shah FA, Mahmood Q, Rashid N, Pervez A, Raja IA, Shah MM. Co-digestion, pretreatment and digester design for enhanced methanogenesis. Renewable and Sustainable Energy Reviews. 2015;**42**:627-642

[14] Sreekrishnan TR, Kohli S, Rana V. Enhancement of biogas production from solid substrates using different techniques—A review. Bioresource Technology. 2004;**95**(1):1-10

[15] Kaparaju P, Serrano M, Angelidaki I. Optimization of biogas production from wheat straw stillage

in UASB reactor. Applied Energy. 2010;**87**(12):3779-3783

[16] Brito AG, Peixoto J, Oliveira JM, Oliveira JA, Costa C, Nogueira R, Rodrigues A. Brewery and winery wastewater treatment: Some focal points of design and operation. In: Utilization of by-Products and Treatment of Waste in the Food Industry. Boston, MA: Springer; 2007. pp. 109-131

[17] Achinas S, Achinas V, Euverink GJ. A technological overview of biogas production from biowaste. Engineering. 2017;**3**(3):299-307

[18] Chan YJ, Chong MF, Law CL. Performance and kinetic evaluation of an integrated anaerobic-aerobic bioreactor in the treatment of palm oil mill effluent. Environmental Technology. 2017;**38**(8):1005-1021

[19] Kainthola J, Kalamdhad AS, Goud VV. Optimization of methane production during anaerobic co-digestion of rice straw and hydrilla verticillata using response surface methodology. Fuel. 2019;**235**:92-99

[20] Aqaneghad M, Moussavi G. Electrochemically enhancement of the anaerobic baffled reactor performance as an appropriate technology for treatment of municipal wastewater in developing countries. Sustainable Environment Research. 2016;**26**(5):203-208

[21] Nilsson M, Griggs D, Visbeck M. Policy: Map the interactions between sustainable development goals. Nature News. 2016;**534**(7607):320

[22] Waage J, Yap C, Bell S, Levy C, Mace G, Pegram T, et al. Governing the UN sustainable development goals: Interactions, infrastructures, and institutions. The Lancet Global Health. 2015;**3**(5):e251-e252

[23] Tiwary A, Williams ID, Pant DC, Kishore VVN. Emerging perspectives on environmental burden minimisation initiatives from anaerobic digestion technologies for community scale biomass valorisation. Renewable and Sustainable Energy Reviews. 2015;**42**:883-901

[24] De Besi M, McCormick K. Towards a bioeconomy in Europe: National, regional and industrial strategies. Sustainability. 2015;**7**(8):10461-10478

[25] Scarlat N, Dallemand JF, Fahl F. Biogas: Developments and perspectives in Europe. Renewable Energy. 2018;**129**:457-472

[26] Liu Y, Chen N, Liu Y, Liu H, Feng C, Li M. Simultaneous removal of nitrate and hydrogen sulfide by autotrophic denitrification in nitrate-contaminated water treatment. Environmental Technology. 2018:1-2

[27] Carley JK, Pasternack GB, Wyrick JR, Barker JR, Bratovich PM, Massa DA, Reedy GD, Johnson TR. Significant decadal channel change 58-67 years post-dam accounting for uncertainty in topographic change detection between contour maps and point cloud models. Geomorphology. 2012;**179**:71-88

[28] Warsinger DM et al. A review of polymeric membranes and processes for potable water reuse. Progress in Polymer Science. 2018;**81**:209-237

[29] Diallo MS. Water treatment by dendrimer-enhanced filtration: Principles and applications. In: Nanotechnology Applications for Clean Water. William Andrew Publishing; 2009. pp. 143-155

[30] Frenkel VS. Membranes in water and wastewater treatment. In: Proceedings of the World Environmental and Water Resources Congress. 2008. pp. 316-324

[31] Hamouda SB, Touati K, Amor MB. Donnan dialysis as membrane

process for nitrate removal from drinking water: Membrane structure effect. Arabian Journal of Chemistry. 2017;**10**:S287-S292

[32] Van Der Bruggen B. Integrated membrane separation processes for recycling of valuable wastewater streams: Nanofiltration, membrane distillation, and membrane crystallizers revisited. Industrial and Engineering Chemistry Research. Jan 2013;**52**(31):10335-10341

[33] Bodzek M. Membrane technologies for the removal of micropollutants in water treatment. In: Advances in Membrane Technologies for Water Treatment: Materials, Processes and Applications. Woodhead Publishing; 2015. pp. 465-517

[34] Shahmansouri A, Bellona C. Nanofiltration technology in water treatment and reuse: Applications and costs. Water Science and Technology. 1 Feb 2015;**71**(3):309-319

[35] Tedesco M, Hamelers HV, Biesheuvel PM. Nernst-Planck transport theory for (reverse) electrodialysis: II. Effect of water transport through ion-exchange membranes. Journal of Membrane Science. 2017;**531**:172-182

[36] Asante-Sackey L, Rathilal D, Tetteh S, Pillay EK. Optimization of donnan dialysis for alum recovery using box behnken design. In: CBU International Conference Proceedings. 2018. pp. 1007-1012

[37] Abdullah WN, Lau WJ, Aziz F, Emadzadeh D, Ismail AF. Performance of nanofiltration-like forward-osmosis membranes for aerobically treated palm oil mill effluent. Chemical Engineering and Technology. 2018;**41**(2):303-312

[38] Guo Y, Qi PS, Liu YZ. A review on advanced treatment of pharmaceutical wastewater. In: IOP Conference Series: Earth and Environmental Science. IOP Publishing; 2017;**63**(1):012025

[39] Mehta BB, Joshi RN, Raval HD. A novel ultra-low energy reverse osmosis membrane modified by chitosan with glutaraldehyde crosslinking. Journal of Applied Polymer Science. 2018;**135**(10):45971-45977

[40] Lin S, Elimelech M. Staged reverse osmosis operation: Configurations, energy efficiency, and application potential. Desalination. 2015;**366**:9-14

[41] Johnson J, Busch M. Engineering aspects of reverse osmosis module design. Desalination and Water Treatment. 2010;**15**(1-3):236-248

[42] Bodzek M, Konieczny K. Membranes in organic micropollutants removal. Current Organic Chemistry. 2018;**22**(11):1070-1102

[43] Dezotti M, Lippel G, Bassin JP. Advanced Biological Processes for Wastewater Treatment: Emerging, Consolidated Technologies and Introduction to Molecular Techniques. Springer; 2017

[44] Pachés M, Martínez-Guijarro R, González-Camejo J, Seco A, Barat R. Selecting the most suitable micro-algae species to treat the effluent from an anaerobic membrane bioreactor. Environmental Technology. 2018:1-10

[45] Roy D, Azaïs A, Benkaraache S, Drogui P, Tyagi RD. Composting leachate: Characterization, treatment, and future perspectives. Reviews in Environmental Science and Biotechnology. 2018

[46] Thiruvenkatachari R, Vigneswaran S, Moon IS. A review on UV/TiO_2 photocatalytic oxidation process. Korean Journal of Chemical Engineering. 2008;**25**(1):64-72

[47] Sievers M. Advanced oxidation processes. In: Treatise on Water Science. 2010;377-408

[48] Bethi B, Sonawane SH, Bhanvase BA, Gumfekar SP. Nanomaterials-based advanced oxidation processes for wastewater treatment: A review. Chemical Engineering and Processing Process Intensification. 2016;**109**:178-189

[49] Ameta S, Ameta R. Advanced oxidation processes for wastewater treatment: Emerging green. Chemical Technology. Academic Press; 2018

[50] Thakur RS, Chaudhary R, Singh C. Fundamentals and applications of the photocatalytic treatment for the removal of industrial organic pollutants and effects of operational parameters: A review. Journal of Renewable and Sustainable Energy. 2010;**2**(4):042701

[51] Demeestere K, Dewulf J, Van Langenhove H. Heterogeneous photocatalysis as an advanced oxidation process for the abatement of chlorinated, monocyclic aromatic and sulfurous volatile organic compounds in air: State of the art. Critical Reviews in Environmental Science and Technology. 2007;**37**(6):489-538

[52] Khaki MR, Shafeeyan MS, Raman AA, Daud WMAW. Application of doped photocatalysts for organic pollutant degradation—A review. Journal of Environmental Management. 2017;**198**:78-94

[53] Diya'Uddeen BH, Daud WMAW, Abdul Aziz AR. Treatment technologies for petroleum refinery effluents: A review. Process Safety and Environmental Protection. 2011;**89**(2):95-105

[54] Hasan DB, Abdul Aziz AR, Daud WMAW. Oxidative mineralisation of petroleum refinery effluent using Fenton-like process. Chemical Engineering Research and Design. 2012;**90**(2):298-307

[55] Kim JR, Kan E. Heterogeneous photocatalytic degradation of sulfamethoxazole in water using a biochar-supported TiO_2 photocatalyst. Journal of Environmental Management. 2016;**180**:94-101

[56] Gaya UI, Abdullah AH. Heterogeneous photocatalytic degradation of organic contaminants over titanium dioxide: A review of fundamentals, progress and problems. Journal of Photochemistry and Photobiology C: Photochemistry Reviews. 2008;**9**(1):1-2

[57] Bhatkhande DS, Pangarkar VG, Beenackers AA. Photocatalytic degradation for environmental applications—A review. Journal of Chemical Technology and Biotechnology: International Research in Process, Environmental & Clean Technology; 2002;**77**(1):102-116

[58] Zhang T, Lin W. Metal-organic frameworks for artificial photosynthesis and photocatalysis. Chemical Society Reviews. 2014;**43**(16):5982-5993

[59] Zhu S, Wang D. Photocatalysis: Basic principles, diverse forms of implementations and emerging scientific opportunities. Advanced Energy Materials. 2017;**7**(23):1700841

[60] Sinai J, Sugarmen C, Fisher U. Adaptation and modification of gas turbines for solar energy applications. In: Proceedings of GT2005 ASME Turbo Expo 2005; 6-9 June 2005; Power for Land Sea and Air: Reno-Tahoe, Nevada, USA. 2005

[61] Sinai J, Sugarmen C, Fisher U. GT2005-68122 balance of plant German aerospace agency. In: Proceedings of GT2005 ASME Turbo Expo 2005; 6-9 June 2005; Power for Land, Sea and Air: Reno-Tahoe, Nevada, USA. 2005

[62] Atputharajah A. Operational challenges of low power hydro plants. In: Handbook of Renewable Energy Technology. Singapore: World Scientific Publishing Co. Pte. Ltd.; 2011;469-483

[63] Xia PJ, Chen FQ. Landscape pattern of the Xiangjiaba hydropower project'disturbed area in 2012 based on the GIS analysis. In: Advanced Materials Research. Trans Tech Publications. 2014;**955**:3854-3858

Chapter 2

Desalination with Renewable Energy: A 24 Hours Operation Solution

Muhammad Wakil Shahzad, Muhammad Burhan, Doskhan Ybyraiymkul and Kim Choon Ng

Abstract

The inevitable escalation in economic development has serious implications on energy and environment nexus. The International Energy Outlook 2016 (IEO2016) predicted that the Non Organization for Economic Cooperation and Development (non-OECD) countries will lead with 71% rise in energy demand in contrast with only 18% in developed countries from 2012 to 2040. In Gulf Cooperation Council countries (GCC) countries, about 50% of primary energy is consumed for cogeneration based power and desalination plants. The desalination capacities are expected to increase fivefold by 2050 and renewable energy application can be one of the solution for sustainable water production. The major bottleneck in commercialization of renewable energy sources is its intermittent nature of supplies specially wind and solar. We proposed solar thermal energy storage to operate desalination system around the clock. Magnesium oxide (MgO) can be utilized as an efficient energy storage system to store solar thermal energy for off period operation. The heat generated by regeneration processes at day time and exothermic adsorption at night can operate desalination cycle 24 h. The operational temperature ranges from 120 to 140°C and energy storage 41–81 kJ/mol. It was successfully demonstrated by experimentation that MgO operated hybrid desalination cycle can achieve highest performance and lowest carbon emission. The proposed cycle can achieve sustainable water production goals.

Keywords: renewable energy, hybrid desalination, low emission desalination, sustainable desalination, energy storage

1. Introduction

The energy demand in Gulf Cooperation Council (GCC) countries is almost doubled in a decades, from 300 TWh in 2000 to 600 TWh in 2012. In GCC countries, the residential and commercial sector energy demand has grown rapidly. The residential sector consumes over 50% due to improved living standards. In GCC countries, per capita primary energy consumption is the highest as compared to other countries in the world as shown in the **Figure 1** [1–3].

The GCC countries also produce substantial amount of CO_2 and it was estimated as 1.2 billion tons of CO_2-equivalent in 2012. The major part of CO_2 emission is related to energy and water production. The GCC countries are the most water

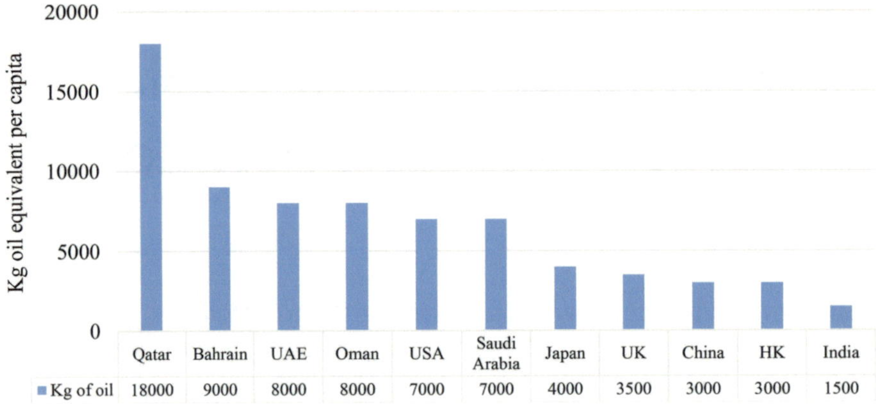

Figure 1.
Per capita oil consumption in different parts of the world.

Country	2010 consumption	2050 consumption	2050 shortage
Bahrain	227	400	380
Kuwait	509	1220	850
Oman	760	1700	1150
Qatar	328	400	175
Saudi Arabia	20,480	27,000	20,100
UAE	3375	3500	3250
Total			

Table 1.
Water consumption and estimated shortage in 2050 in GCC countries.

scariest in the world due to dry environmental conditions and recently it became even worst due to population increase and GDP growth. It is estimated that by 2050, the shortage of water supply can be as high as 77%. **Table 1** shows the water consumption and shortage in million cubic meter per year scenario in all GCC countries by 2050. It can be noticed the large gap in water demand and supply cannot be filled by renewable and ground water sources. The non-renewable such as desalination is the only source for future water supplies in GCC countries [1–3].

Today, all desalination processes are energy intensive and consume primary energy in the range of 6–10 kWh/m^3. The inefficiency of desalination processes, 10–13% of thermodynamic limit, requires not only more energy but they also emit enormous CO_2 [4–7]. For future sustainability, one feasible option is to utilize renewable energy such as wind and solar. The renewable technologies have drastically developed and their economics are greatly improved in recent years, and GCC countries have great potential to exploit the renewable energies potential such as solar and wind. The GCC countries government announced mega investment plans to invest in renewable energy sectors to meet the increasing demand of electricity as shown in **Figure 2** [8].

The resettlement of energy mix in GCC countries needs a comprehensive plan for contractor as well as operator companies. One of the major challenges in renewable energies development is its intermittent nature. Currently they are only employed to cope the peak load typically during office hours. The one of the solution to overcome intermittent supply is the energy storage and there are two methods namely, battery storage and thermal heat storage. In terms of battery storage, the efficiency is very low, typically 8–10% in field operation due to

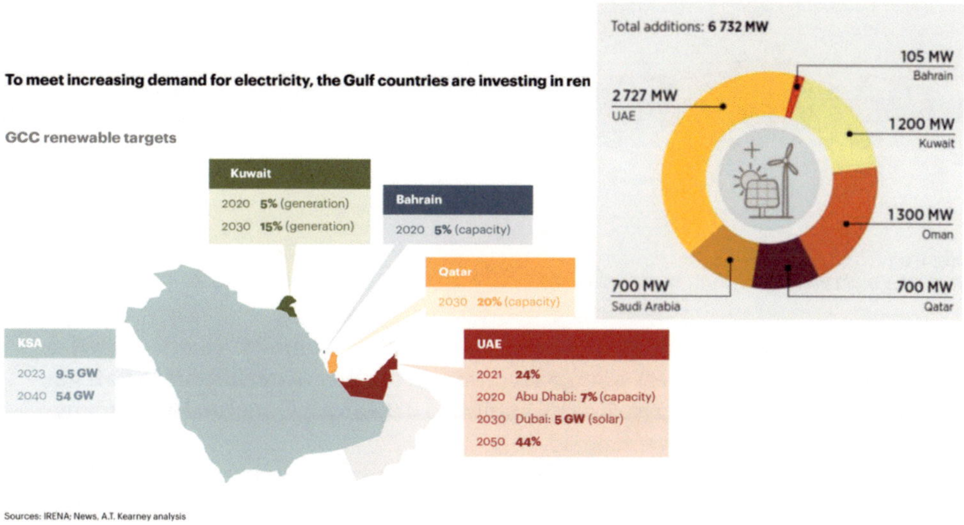

Figure 2.
GCC countries renewable energy development plan by 2030.

Materials	Heat storage method		Heat storage density (GJ/m^3)	Heat charging temperature (°C)
Co_3O_4	Thermochemical materials (TCM)	Inorganic oxides	5.0	925
CaO			4.5	550
MgO			3.4	350
$MgSO_4$		Anhydrate	2.6	125
Silica gel		Adsorbate	0.8	85
Zeolite			0.6	220
Paraffin	Phase change materials		0.2	60
Water	Sensible		0.2	0–100

Table 2.
Comparison of different thermal energy storage materials.

efficiencies involved from one form to other form conversions. On the other hand, direct thermal storage and utilization efficiency is significantly high due to same form of energy utilization without conversion into different forms. There are three major technologies utilize different methods to store solar energy. The comparison of different heat storage materials is summarized in **Table 2** [9–11].

2. Solar thermal energy storage and desalination application

Thermochemical materials (TCM) have many advantages over the other materials such as high heat storage density and low heat leak. Once the reactant leave the thermochemical materials, the enthalpy remains same and it help to achieve the state of energy charging. Subsequently, the discharged energy is utilized while the material remains stable. In the past, a lot of studies were carried out on heat pump using different TCM materials [12–14]. The selection of TCM materials for different application is based on many elements such as (i) heat storage temperature, (ii) heat releasing temperature,

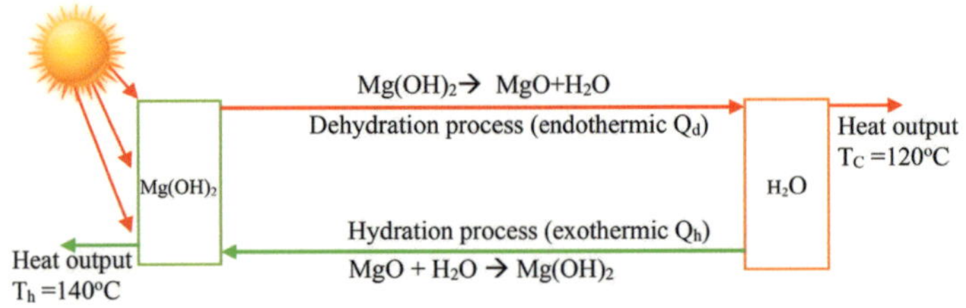

Figure 3.
MgO thermal energy storage system operation.

(iii) heat storage density, and (iv) material stability. The magnesium oxide (MgO) is most suitable for thermal heat storage as compared to other materials due to its high density and stability. Many researchers published data on MgO thermal heat storage and its performance improvement [15, 16].

The dry MgO reacts with water (hydration) to become hydrated $Mg(OH)_2$. The hydration is a exothermic reaction and generates 81 kJ/mol. During dehydration of $Mg(OH)_2$ it becomes MgO through a reverse process at 350–500°C from solar collectors and high temperature vapor are utilized as a heat source. It can be noticed that MgO as an energy storage material produce heat during day as well as night time. The hydration process at night and dehydration process at day with solar energy can produce sufficient heat energy to operate the desalination cycle.

The principle of this heat pump is shown in **Figure 3**. The heat pump consists of a magnesium oxide reactor and a water reservoir. In heat storage mode magnesium hydroxide ($Mg(OH)_2$) is dehydrated by surplus heat (Q_d) at T_d from sun. The generated vapors are condensed at the reservoir at T_C and the condensation heat (Q_d) of the vapor is used for desalination cycle at day time. The hydration of magnesium oxide proceeds in the reactor by introducing the vapor, and a hydration heat output (Q_h) at T_h is generated to operate desalination cycle at night. Thermal drivability, which does not require mechanical work, is one of the advantages of the heat pump. The environmentally friendly and economical nature of the reactants is also advantageous. This type of heat pump is able to store heat at around 350°C through $Mg(OH)_2$ dehydration and to transfer stored heat at temperatures between 110 and 150°C through MgO hydration. The solar thermal energy storage and 24 h delivery around 100°C is best suitable for sustainable desalination processes [17–19].

The renewable energy (RE) driven desalination processes are already commercialized but at low scale due to some operational complexity. **Table 3** summarized

Plant name	Location	Technology	Capacity	Energy source	Cost* (US$/m^3)
Kimolos	Greece	MED	2000	Geothermal	2.5–3
Keio university	Japan	MED	100	Solar thermal	
PSA	Spain	MED	72	CSP	
Ydriada	Greece	RO	80	Wind	2–6
Morocco	Morocco	RO	20	PV	2–5
Oyster	Scotland	RO	—	Wave energy	3–5
KAUST**	Saudi Arabia	MEDAD hybrid	10	Solar thermal	0.5

**Cost is estimated based on plant capacity more than 1000 m^3/day [20].*
***Refs. [21–29].*

Table 3.
RE driven desalination technologies and water cost.

the major renewable desalination plants operated in the world and estimated cost of water production.

It can be noticed that thermal desalination processes are most favorable option with solar thermal energy operation. The most efficient thermally driven multi effect desalination (MED) system recently investigated to overcome its conventional operational limitations. The numbers of stages in a MED is controlled by top brine temperature (TBT) and lower brine temperature (LBT). The TBT typically 70°C is restricted by soft scaling components in the feed water and the LBT is controlled by ambient condition due to water cooled condenser to condense the last stage vapors. The MED system can be more efficient if these two limitations cab be removed to increase number of recoveries. Researchers found that TBT can be increased to 125°C by inducing nano-filtration (NF) prior to introduction the feed into the system. This NF process helps to remove soft scaling components and prevent scaling and fouling on the tubes of evaporators. The inter stage temperature and the last stage operating temperature limitations can be overcome by hybridization with adsorption cycle. AD cycle can operate below ambient conditions typically as low as 5°C due high affinity of water vapors of adsorbent (silica gel). MED last stage temperature can be lower down to 5°C by introducing the AD at downstream. The proposed hybrid MEDAD system with TES will be the best choice for sustainable water supplies.

3. Proposed system operation

The detailed schematic of TES driven hybrid MEDAD desalination cycle is shown in **Figure 4**. During day time operation, solar heat is supplied to the hydrated $Mg(OH)_2$ at 300–400°C and regenerated vapor condensation heat at 120–150°C is utilized to operate desalination cycle. At night time, the hydration of MgO generates sufficient heat due to exothermic reaction that is supplied to the desalination cycle to continue the operation. It can be noticed that with MgO energy storage system, thermal desalination can be operated for 24 h using solar energy.

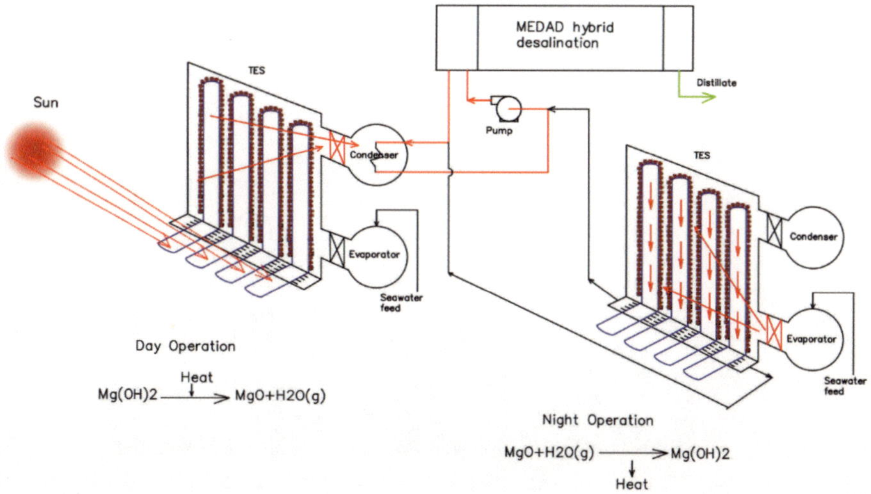

Figure 4.
Proposed TES driven hybrid MEDAD desalination cycle.

4. Experimentation

An experimental system was designed and installed to test workability of proposed concept. **Figure 5** shows the temperature profiles of MEDAD effects at heat source of 45°C. The pilot was tested at different temperatures to investigate the performance. **Figure 6** shows the hybrid MEDAD system effects temperatures at different heat source temperatures. The system performed well as per designed 3–4°C inter-effect temperature difference. Similarly, **Figure 7** shows the corresponding saturation pressures.

Figure 8 shows the water production profiles of MED effects, AD condenser and total production at 45°C heat source temperature. The summary of water production presented in **Figure 9** at different heat source temperatures. It can be seen that at higher temperature the water production is also higher and it drop due to drop in heat capacity. The system is designed for 45°C operational temperature but it performed well at off-design conditions. It shows the robustness of the thermally driven desalination systems.

The thermal energy consumed is shown in **Figure 10**. It can be noticed that at higher heat input temperature the energy consumed by the system is also higher. It is mainly due to the higher temperature difference between heat inlet and out

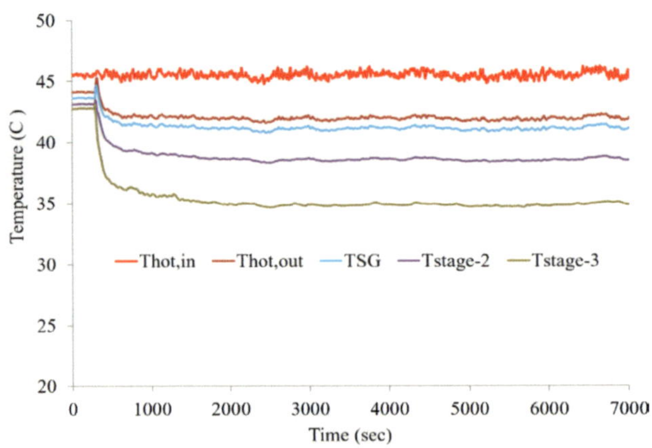

Figure 5.
Hybrid MEDAD temperature profiles at 45°C heat source.

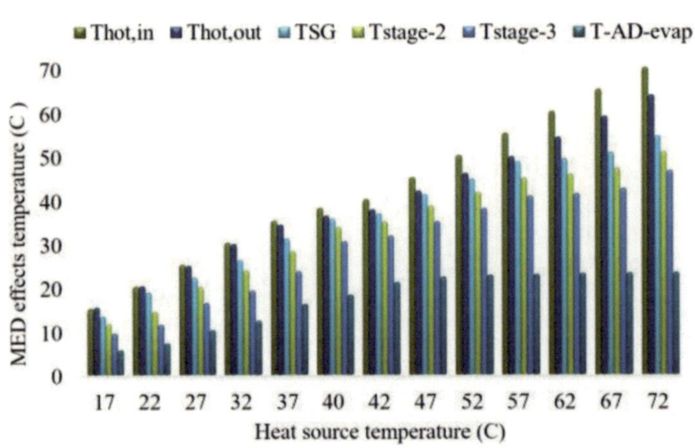

Figure 6.
Hybrid MEDAD inter-effect temperatures at different heat source (reproduce with author's permission [30].

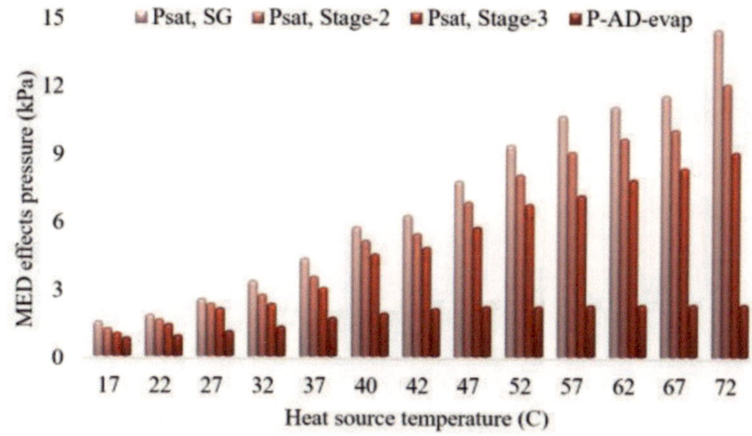

Figure 7.
Hybrid MEDAD inter-effect pressures at different heat source (reproduce with author's permission [30]).

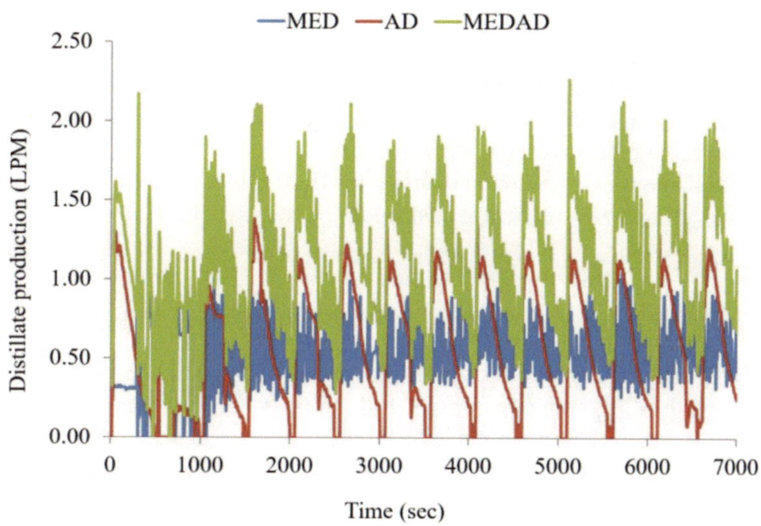

Figure 8.
Hybrid MEDAD water production profiles at 45°C heat source (reproduce with author's permission [30]).

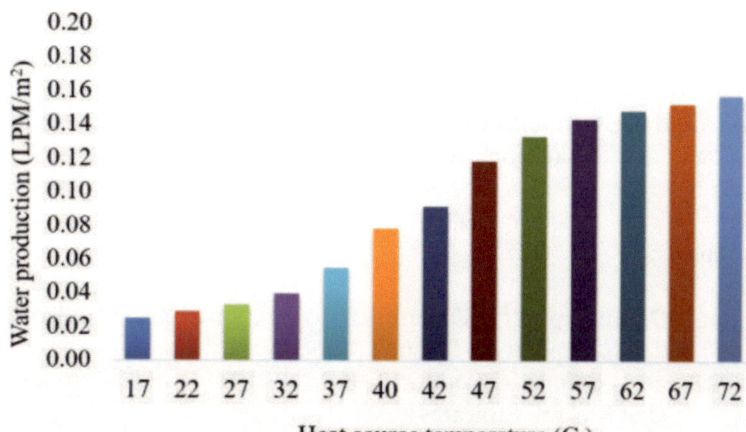

Figure 9.
Hybrid MEDAD water production at different heat source temperature (reproduce with author's permission [30]).

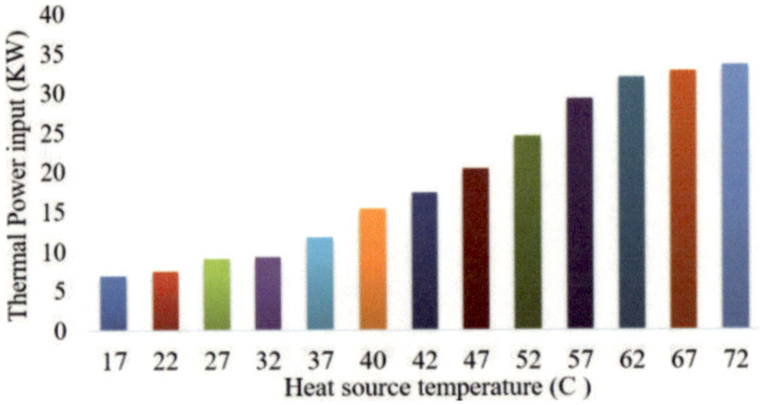

Figure 10.
Hybrid MEDAD thermal energy input at different heat source temperature (reproduce with author's permission [30]).

temperatures. The interesting trend was noticed at below 25°C where heat input showed negative value. It is because the heat was scavenged from the ambient. The system was operating below ambient conditions due to adsorption cycle hybridization that allows last effects to operate as low as 5°C.

The successful experimentation of hybrid MEDAD cycle proved the workability of TES + MEDAD system for future sustainable water supplies.

5. Conclusion

Thermal energy storage based hybrid desalination system is proposed for 24 h operation. MgO has high energy density and stability for long term operation. The proposed TES + MEDAD hybrid cycle has highest performance. The superiority of MEDAD cycle has been successfully demonstrated pilot as compared to conventional MED system by improving water production to twofold as same heat source temperature. The proposed combination is estimated to have highest performance to achieve sustainability goals. These innovative solutions will help to save energy and protect environment.

Acknowledgements

Authors would like to thanks to KAUT and NUS for this study. The data is reproduced by PI and the author's permission [30].

Nomenclature

OECD	Organization for Economic Cooperation and Development
GCC	Gulf Cooperation Council
GDP	gross domestic product
UAE	United Arab Emirates
TCM	thermochemical materials
RE	renewable energy
MED	multi effect desalination

MSF	multi stage flash
SWRO	seawater reverse osmosis
AD	adsorption
TES	thermal energy storage
CSP	concentrated solar photovoltaic
TBT	top brine temperature
LBT	lower brine temperature
LPM	liter per minute

Author details

Muhammad Wakil Shahzad*, Muhammad Burhan, Doskhan Ybyraiymkul and Kim Choon Ng
Water Desalination and Reuse Centre (WDRC), Biological and Environmental Science and Engineering (BESE), King Abdullah University of Science and Technology, Saudi Arabia

*Address all correspondence to: muhammad.shahzad@kaust.edu.sa

References

[1] Energy Use (kg of Oil Equivalent Per Capita). World Bank Data. Available from: https://data.worldbank.org/indicator/EG.USE.PCAP.KG.OE [Accessed: Jan 02 2019]

[2] Renewable Energy Market Analysis. The GCC region. The International Renewable Energy Agency (IRENA) Report 2016

[3] Parmigiani L. Water and Energy in the GCC: Securing Scarce Water in Oil-Rich Countries. The Institut français des relations internationales (Ifri) Report 2015. ISBN: 978-2-36567-442-3

[4] Shahzad MW, Burhan M, Ng KC. A standard primary energy approach for comparing desalination processes. Nature Clean Water. 2019;**1**:1-7

[5] Shahzad MW, Burhan M, Ybyraimkul D, Ng KC. Desalination processes efficiency and future roadmap. Entropy. 2019;**21**(1):84

[6] Shahzad MW, Burhan M, Son HS, Seung Jin O, Ng KC. Desalination processes evaluation at common platform: A universal performance ratio (UPR) method. Applied Thermal Engineering. 2018;**134**:62-67

[7] Ng KC, Shahzad MW, Son HS, Hamed OA. An exergy approach to efficiency evaluation of desalination. Applied Physics Letters. 2017;**110**:184101. DOI: 10.1063/1.4982628

[8] Partner, Kearney AT. Why oil-rich gulf countries need to invest in renewable energy. World Economic Forum. Available from: https://www.weforum.org/agenda/2017/05/why-oil-rich-gulf-countries-need-to-invest-in-renewable-energy/ [Accessed: Jan 15 2019]

[9] Yim T, Kim HS, Lee JY. Cyclic assessment of magnesium oxide with additives as a thermochemical material to improve the mechanical strength and chemical reaction. Energies. 2018;**11**:2366

[10] Singh A, Tescari S, Lantin G, Agrafiotis C, Roeb M, Sattler C. Solar thermochemical heat storage via the Co_3O/CoO looping cycle: Storage reactor modelling and experimental validation. Solar Energy. 2017;**144**:453-465

[11] Fernandes MS, Brites GJVN, Costa JJ, Gaspar AR, Costa VAF. Modeling and parametric analysis of an adsorber unit for thermal energy storage. Energy. 2016;**102**:83-94

[12] Cabeza LF, Martorell I, Miró L, Fernández AI, Barreneche C. Advances in Thermal Energy Storage Systems: Methods and Applications, Advances in Thermal Energy Storage Systems. Woodhead Publishing. ISBN: 978-1-78242-088-0

[13] Alva G, Lin Y, Fang G. An overview of thermal energy storage systems. Energy. 2018;**144**:341-378

[14] Thomas JJ, Musso S, Prestini I. Kinetics and activation energy of magnesium oxide hydration. The American Ceramic Society Journal. 2014;**97**:275-282

[15] Mastronardo E, Kato Y, Bonaccorsi L, Piperopoulos E, Milone C. Thermochemical storage of middle temperature wasted heat by functionalized C/Mg(OH) hybrid materials. Energies. 2017;**10**:70

[16] Mastronardo E, Kato Y, Bonaccorsi L. *Piperopoulos elpida*, efficiency improvement of heat storage materials for $MgO/H_2O/Mg(OH)_2$ chemical heat pumps. Applied Energy. 2016;**162**:31-39

[17] Kato Y, Takahashi F, Sekiguchi T, Ryu J. Study on medium temperature chemical

heat storage using mixed hydroxides. International Journal of Refrigeration. 2009;**32**:661-666

[18] Kato Y, Nakahata J, Yoshizawa Y. Durability characteristics of the hydration magnesium oxide under repetitive reaction. Journal of Material Sciences. 1999;**34**:475-480

[19] Kato Y, Yamashita N, Kobayashi K, Yoshizawa Y. Kinetic study of the hydration of magnesium oxide for a chemical heat pump. Applied Thermal Engineering. 1996;**16**:853-862

[20] Alkaisi A, Mossad R, Sharifian-Barforoush A. A review of the water desalination systems integrated with renewable energy. Energy Procedia. 2017;**110**:268-274

[21] Shahzad MW, Burhan M, Ang L, Ng KC. Energy-water-environment nexus underpinning future desalination sustainability. Desalination. 2017;**413**:52-64

[22] Shahzad MW, Ng KC. An improved multi-evaporator adsorption desalination cycle for GCC countries. Energy Technology. 2017. DOI: 10.1002/ente.201700061

[23] Shahzad MW, Ng KC, Thu K. Future sustainable desalination using waste heat: Kudos to thermodynamic synergy. Environmental Science: Water Research & Technology. 2016;**2**:206-212

[24] Thu K, Kim Y-D, Shahzad MW, Saththasivam J, Ng KC. Performance investigation of an advanced multi-effect adsorption desalination (MEAD) cycle. Applied Energy. 2015;**159**:469-477

[25] Shahzad MW, Thu K, Kim Y-d, Ng KC. An experimental investigation on MEDAD hybrid desalination cycle. Applied Energy. 2015;**148**:273-281

[26] Ng KC, Thu K, Oh SJ, Ang L, Shahzad MW, Ismail AB. Recent developments in thermally-driven seawater desalination: Energy efficiency improvement by hybridization of the MED and AD cycles. Desalination. 2015;**356**:255-270

[27] Shahzad MW, Ng KC, Thu K, Saha BB, Chun WG. Multi effect desalination and adsorption desalination (MEDAD): A hybrid desalination method. Applied Thermal Engineering. 2014;**72**:289-297

[28] Ng KC, Thu K, Shahzad MW, Gee CW. Progress of adsorption cycle and its hybrids with conventional multi-effect desalination processes. IDA Journal of Desalination and Water Reuse. 2016;**6**:1, 44-56

[29] Shahzad MW, Thu K, Saha BB, Ng KC. An emerging hybrid multi-effect adsorption desalination system. Evergreen Joint Journal of Novel Carbon Resource Sciences & Green Asia Strategy. 2014;**01**:02, 30-36

[30] Muhammad Wakil Shahzad, The hybrid multi-effect desalination (MED) and the adsorption (AD) cycle for desalination, Doctoral Thesis, National University of Singapore. 2013

Chapter 3

Nonconventional Wastewater Treatment for the Degradation of Fuel Oxygenated (MTBE, ETBE, and TAME)

Zenaida Guerra Que, José Gilberto Torres Torres, Ignacio Cuauhtémoc López, Juan C. Arévalo Pérez, Adrian Cervantes Uribe, Hermicenda Pérez Vidal, Alejandra E. Espinosa de los Monteros Reyna, José G. Pacheco Sosa, María A. Lunagómez Rocha and Cecilia Sánchez Trinidad

Abstract

Catalytic wet air oxidation (CWAO) is a nonconventional wastewater treatment, consisting of oxygen pressure releasing inside a reactor in order to degrade organic compounds dissolved in water, using a solid catalyst in the presence of an activated O_2 species, usually at temperatures ranges of 125–250°C and pressures of 10–50 bar. CWAO can reduce operating costs of conventional treatment due to the use of ideal catalyst that is able to improve reaction conditions at temperatures and pressures as mild as possible, simultaneously setting high catalytic activity and long-term stability of heterogeneous catalysts. Oxygenated fuels are gasoline additives in reformulated gasoline and oxyfuels. In the beginning, they provided an alternative solution of environmental problems, such as greenhouse gas emissions and octane enhancement, caused by fossil fuel use. The oxygenated fuels frequently used are methyl tert-butyl ether (MTBE), ethyl tert-butyl ether (ETBE), and tert-amyl methyl ether (TAME). However, there is environmental impact from oxygenated fuel hydrocarbons related to widespread contamination of groundwater and other natural waters. Our research group developed a wide study in order to evaluate several catalysts (Ru, Au, Cu, and Ag supported on Al_2O_3, Al_2O_3-CeO_2, and TiO_2-CeO_2) and to obtain the best for the efficiency of the oxidation process.

Keywords: nanoparticles, wastewater, CWAO, fuel oxygenated

1. Introduction

1.1 Oxygenated fuels

Oxygenated fuels are oxygen-rich compounds such as alcoholic and ether fuels that act as gasoline additives in reformulated gasoline and oxyfuels. Oxygenates can be blended into gasoline in two forms: alcohols (such as methanol or ethanol) or ethers. They have potential to provide an alternative solution of environmental problems caused by fossil fuel use. The oxygenated ether fuels used more frequently are methyl tert-butyl ether (MTBE), ethyl tert-butyl ether (ETBE), and tert-amyl methyl ether (TAME), although the first fuel oxygenate used in reformulation was MTBE. The use of MTBE as an octane enhancer in the United States began in 1979 [1, 2].

MTBE is a compound (chemical formula $C_5H_{12}O$) that is synthetized by the chemical reaction of methanol and isobutylene, and it is almost exclusively used as a fuel additive in motor gasoline. Although ETBE and TAME have only secondary importance in production industry, recently they have become more prevalent, and they can be used instead of MTBE. In France, MTBE has been partially replaced by the ETBE since 1990 or as part of a "binary or ternary mixture" with MTBE [3–5].

ETBE (chemical formula $C_6H_{14}O$) is considered an attractive octane enhancer as it presents a lower reid vapor pressure (RVP) mixture and lower water solubility than MTBE and ethanol. The azeotropic mixture of ETBE with ethanol takes down the volatility of ethanol making it suitable as an additive for automatic gasoline [6, 7].

Finally, TAME (chemical formula $C_6H_{14}O$) was considered as an oxygenated fuel until the 1990s despite its octane rating lower than other oxygenated fuels; besides it is very soluble with other ethers, and it is highly soluble in water (12 g/l) [8, 9].

1.2 Importance of oxygenated fuels in petroleum industry

The leading produced fuel in the world is gasoline. Because of its widespread use and the fact that it is composed of that fraction of crude oil with lower boiling points, gasoline is the single largest source of volatile hydrocarbons to the environment. Motor gasoline comes in various blends with properties that affect engine performance. All motor gasolines are made of relatively volatile components of crude oil. Other fuels include distillate fuel oil (diesel fuel and heating oil), jet fuel, residual fuel oil, kerosene, aviation gasoline, and petroleum coke. In the petroleum refining process, heat distillation is used first to separate different hydrocarbon components. The lighter products are liquefied petroleum gases and gasoline, whereas the heavier products include heavy gas oils. Liquefied petroleum gases include ethane, ethylene, propane, propylene, n-butane, butylenes, and isobutane. Internal combustion engines of high compression ratio require gasoline with octane ratings that are sufficiently high to ensure efficient combustion [2, 10, 11].

Gasolines need additives that increase their octane rating so they can decrease their self-knock capacity, increasing their resistance to compression, and finally improve the quality of gasoline. An economical way of achieving these properties has been the use of anti-knock additives, such as tetraethyl and tetramethyl lead at concentrations up to 0.84 g/l. With the phasing out of lead from gasoline because it was increasingly recognized that lead is toxic and non-biodegradable, oxygenated fuel become a better alternative for gasoline additive, instead of lead. Oxygenated fuels act as octane enhancers, bringing the additional benefit of making gasoline burn almost completely. Actually, using oxygenated fuels in internal combustion engine leads to a reduction of greenhouse gas (GHG) emissions, compared to gasoline because they burn cleaner than regular gasoline and produce lesser carbon monoxide (CO) and nitrogen oxides (NOx) and they reduce emissions of unburned hydrocarbons. Furthermore, using oxygenated fuels in an internal combustion engine (ICE) provides an alternative to conventional fuels that can solve many environmental problems [6, 12, 13].

In the United States, air quality regulations placed on automobile exhaust gases have forced dramatic changes in gasoline formulations. By 1990, the Clean Air Act Amendments (CAAA) required additives, such as MTBE at 15% and ethanol, to be blended in gasoline in some metropolitan areas, heavily polluted by carbon monoxide, and to reduce carbon monoxide and ozone concentrations [14].

1.3 Environmental impact of oxygenated fuels

Despite providing better conditions in terms of fuel quality, an environmental impact from oxygenated fuel hydrocarbons related to widespread contamination of groundwater and other natural waters exists. The distribution and storage of crude oil and refined products result in releases of significant amounts of hydrocarbons to the atmosphere, surface waters, soils, and groundwater. Groundwater contamination by crude oil, and other petroleum-based liquids, is a particularly widespread problem. In Mexico, the agency in charge of producing and distributing fuels derived from petroleum distillation, such as gasoline, diesel, fuel oil, diesel, and LP gas, is Petróleos Mexicanos (PEMEX). The retail distribution of gasoline and diesel is carried out by service stations (gas stations). One of the environmental risks that involves the handling of these stations is spills or leaks of fuels, which cause the contamination of the sites where the storage tanks are located [15].

Unfortunately, these oxygenates have high water solubility and high volatility, causing a high concentration of oxygenated fuel in the environment, air, and water. Another important problem happens when oxygenated fuel is accumulated in the groundwater due to it not absorbing appreciably to soil and undergoing only in slow biodegradation compared to the benzene, toluene, ethylbenzene, and the xylenes (BTEX) in gasoline. The relatively recalcitrant nature of oxygenated fuel to microbial attack makes them persistent, due to them being refractory to the biological treatments. Because of the chemical structure of these, oxygenated fuel hinders their natural biodegradation, which contains a combination of two biorecalcitrant organic functional groups: the ether bond and tertiary carbon atom. These are the reasons why water supplies close to the production sites of MTBE, ETBE, and TAME or near underground petroleum storage tanks and fuelling stations are often contaminated by large amounts of these compounds [16, 17].

There have been extensive occurrences of groundwater contamination by MTBE in the United States because of its prevailing use. In a sampling study of 1208 domestic wells in the United States, MTBE was the most frequently detected fuel oxygenate and the eighth most commonly detected VOC. Perhaps the most publicized case of MTBE contamination of groundwater is the one involving public water supply wells in Santa Monica, California. In August 1995, the city of Santa Monica discovered MTBE in wells used for drinking water supply through routine analytical testing of well water [18, 19].

MTBE has been detected in snow, storm water, surface water (streams, rivers, and reservoirs), groundwater, and drinking water, based on limited surveillance operations conducted in the United States. MTBE concentrations found in storm water ranged from 0.02 to 8.7 μg/l, with a median value of less than 1.0 μg/l. In streams, rivers, and reservoirs, the detection range was 0.2–30 μg/l, and the range of median values in several studies was from 0.24 to 7.75 μg/l [5, 19].

In fact, the US Environmental Protection Agency (USEPA) included MTBE in its Contaminant Candidate List. MTBE in drinking water is carcinogenic for humans and animals. USEPA established a drinking water health advisory of 20–40 μg/l MTBE in December 1997, because it is hazardous to human health (US Environmental Protection Agency, 1997).

Although alternative ether oxygenates are detected less frequently than MTBE, these alternative oxygenates show future groundwater contaminations similar to MTBE if they are not under control.

The toxicokinetic data on MTBE in people come mainly from controlled studies of healthy adult volunteers and in a population exposed to oxygenated gasoline. MTBE quickly passes into circulation after inhalation exposure. In healthy volunteers exposed to inhalation, MTBE kinetics was linear up to concentrations of 268 mg/m^3 (75 ppm). It was measured in the blood and urine of people exposed to tertiary butyl alcohol, metabolic MTBE. The maximum blood concentrations of tertiary butyl alcohol were 17.2–1144 μg/m^3 and 7.8–925 μg/m^3, respectively, in people exposed between 5.0 and 178.5 mg/m^3 (1.4–50 ppm) of MTBE. Based on a single-behavior model, rapid (36–90 min) and slow (19 h) components of MTBE half-life were identified (41). Following the introduction of two separate fuel programs in the United States, which require the use of gasoline oxygenation products, consumers in some areas have complained of acute health disorders, such as headaches, irritation of the eyes and nose, cough, nausea, dizziness, and disorientation. The acute experimental toxicity (CL_{50}) of MTBE in fish, amphibians, and crustaceans is greater than 100 mg/l.

WAO consists of an oxidation in aqueous medium at high temperatures using pure oxygen or high pressure air as an oxidant to maintain the liquid phase. The pressures used and reported in the literature range from 20 to 200 bar and temperature between 150 and 350°C, making this process highly expensive for industrial application. The use of catalysts allows reducing the temperature and pressure conditions for oxidation and even increasing the selectivity toward CO_2. That is why we employ catalytic wet air oxidation, instead of WAO. CWAO of MTBE and oxygenated fuels of gasoline as ETBE and TAME is a nonconventional treatment for degradation of organic compounds in aqueous medium. Our research group developed a wide study in order to evaluate several catalysts and to know what the best are for the efficiency of oxidation process and the total mineralization of pollutants into CO_2 and H_2O.

2. Synthesis, characterization, and catalytic activity of noble (Ru, Au, Ag) and based (Cu) metal nanoparticles supported applied in the CWAO of fuel oxygenated

2.1 Synthesis of Al_2O_3 and Al_2O_3-CeO_2 by wet impregnation and precursor calcination

The synthesis methods occupied for the production of supported catalysts include different techniques or procedures based on a phenomenon of precipitation, chemical adsorption, hydrolysis-polymerization, etc. These methods can synthesize supported catalysts in a single step, or in two steps, that is, both the precursor salt of the support and the active phase are added in the reaction mixture in a single step; otherwise, in sequential or two steps, first the support is synthesized, usually an oxide, and then the active phase, usually a metal, is prepared by some other specific method, expecting all the metal to be added and adsorbed on the support, without metal loss and with a high metal dispersion [20, 21].

These methods determine important properties such as homogeneous metal dispersion, high specific surface area, adequate acidity/basicity ratio, metal-support interaction, and generation of structural defects, for example, oxygen vacancies and reducibility; an improvement in the catalytic performance is concluded owing to the development of these properties in the synthesized catalysts [22, 23]. So in this, study we evaluated different synthesis methods for the catalysts tested in CWAO from oxygenated fuels.

The γ-alumina was obtained by the calcination of boehmite Catapal-B (AIO(OH)), in this process an amount of boehmite (AIO(OH)) is deposited in a fixed-bed quartz reactor in which a continuous flow of air of 1 cm^3/s is passed, and then the calcination is carried out, at a temperature of 650°C for 4 h.

Wet impregnation method was used to prepare the Al_2O_3-CeO_2 support. Ceria is incorporated into the boehmite (AIO(OH)) with an aqueous solution of $Ce(NO_3)_3.6H_2O$ (necessary amount of salt to obtain 1, 3, 5, 7.5, and 10% weight) in 100 ml of distilled water. The precursor solution of ceria is previously deposited in a ball flask, the boehmite is added to this solution and left to stir for 3 h in a rotary evaporator, and then the solution is dried with constant agitation at 60°C to evaporate the water excess. After impregnation, the obtained solid sample was dried at 120°C for about 16 h and calcined at a temperature of 650°C in air flow of 1 cm^3/s for 4 h. The CeO_2 support was obtained commercially.

2.2 Synthesis of noble and base metal catalysts by wet impregnation

The solid catalysts reported in the literature that are used in the oxidation of water pollutants can be classified into four groups: supported metal oxides, unsupported metal oxides, supported metals, and mixtures of noble metals and metal oxides. The type of supported metal is composed mainly of noble and base metals. These are also very important to influence catalyst activity. Noble metals such as Ag, Au, Ru, Pd, Rh, and Pt are very active elements for oxidation reactions; they reveal high activities and excellence stability; however, their high cost and limited availability can decrease their applicability. Base catalysts such as Ni and Cu are more interesting systems, and a lot of research is being done to improve their stability because by having a lower cost, compared to noble metals, they are an economical option; they are also active, but less stable, and suffer from carbon deposit and metal leaching [24, 25].

Cu catalysts supported on Al_2O_3 were synthesized by wet impregnation in a single step. A calculated amount of copper nitrate to obtain a concentration by weight of 5, 10 and 15wt% in copper plus an adequate amount of boehmite Catapal-B were dissolved in 100 ml of water; then the solution was adjusted to a pH of 1, with the addition of a drop of HNO_3, and stirred for 4 h, regulating the temperature from 70 to 90°C. After impregnation, the obtained solid sample was dried at 120°C for about 12 h and calcined at a temperature of 400°C in airflow of 1 cm^3/s for 4 h. Cu (5wt%)/Al_2O_3, Cu (10wt%)/Al_2O_3, and Cu (15wt%)/Al_2O_3 are the monometallic Cu catalysts supported in alumina, synthesized by wet impregnation method in a single step, which later we will name as Cu_5AlIH, Cu_{10}AlIH, and Cu_{15}AlIH.

Copper catalysts supported on Al_2O_3 were also synthesized by sol-gel in a single step. An aqueous solution of 10 ml of aluminum trisecbutoxide ($[C_2H_5CH(CH_3)O]_3Al$), 97% aldrich, d = 0.96 g/mol with 4 g of urea and copper nitrate adequate amount in grams for the percentages of 5, 10, and 15wt% in 1-butanol was progressively added, between 70 and 90°C to a mixture of water and butanol, under constant stirring. After 24 h reflux at 70°C, the resulting pseudo-gel was dried in in a rotating evaporator at 120°C for 12 h and then calcined at 400°C for 4 h. It is worth mentioning that a catalyst was prepared with a pyrrolidine additive instead of urea, exclusively with the same quantities of reagents as the 15wt% in Cu. The synthesized monometallic catalysts will be named as Cu_5AlSG, Cu_{10}AlSG, Cu_{15}AlSG, and Cu_{15}AlSGp.

Finally, the monometallic Cu/Al_2O_3 catalysts were synthesized by wet impregnation with urea and with a concentration by weight of 5, 10, and 15wt% of the metal. A calculated amount of boehmite Catapal-B was dissolved in 300 ml of deionized water; then the solution was adjusted to a pH of 3 with the addition of 1 ml of HNO_3 and stirred for 2 h. After that, 200 ml of a solution of cupric nitrate [$Cu(NO_3)_2$ ½H_2O] and urea is added dropwise to the solution of boehmite Catapal-B, regulating the temperature from 70 to 90°C. After impregnation, the obtained solid sample was washed three times with hot water and dried at 120°C, and finally it was calcined at a temperature of 400°C for 4 h. It should be noted that a catalyst was prepared with a pyrrolidine additive instead of urea, exclusively with the same amounts of reagents as the 15wt% in Cu. The synthesized monometallic catalysts will be named as Cu_5AlIHU, Cu_{10}AlIHU, Cu_{15}AlIHU, and Cu_{15}AlIHp.

Ru-supported catalysts were prepared by wet impregnation method of Al_2O_3 and Al_2O_3-CeO_2 supports aggregating the appropriated amounts of an aqueous solution containing $RuCl_3$* XH_2O to obtain a nominal concentration of 2wt% of Ru, adding 100 ml of hydrochloric acid 0.1 M. First Al_2O_3 and Al_2O_3-CeO_2 (1.0, 3.0, 5.0, 7.5, and 10wt% of Ce) support was wetted by distilled water in a beaker in order to have high dispersion and to maximize the mass transfer of added metal salt ($RuCl_3$* XH_2O) on the surface and the pores of the catalyst. The resulting solution is stirred for 1 h; after that it is heated at 60°C. The samples were dried at 120°C for 24 h and then calcined under air flow (60 ml/min) at 650°C for 4 h, with a heat rate of 2°C/min. Finally, the catalysts were reduced under H_2 (60 ml/min) at 400°C for 5 h, with a heat rate of 2°C/min. The synthesized monometallic catalysts will be named as RuAlIH, RuAlCe_1IH, RuAlCe_3IH, RuAlCe_5IH, RuAl$Ce_{7.5}$IH, and RuAlCe_{10}IH.

2.3 Synthesis of noble and base metal catalysts by deposition-precipitation

Deposition of gold into the modified supports was carried out by the method of deposition-precipitation using urea according to the procedure described below. Support powder (Al_2O_3, CeO_2, Al_2O_3-CeO_2 (1wt%), Al_2O_3-CeO_2 (5wt%), Al_2O_3-CeO_2 (10wt%) was first dispersed in distilled water. The temperature of the suspension was kept constant at 80°C and agitated with a magnetic stirrer. Secondly, the requisite quantity of chloroauric acid ($HAuCl_4$) solution was added to the suspension, and the temperature was let to stabilize. Thirdly, 2.33 g of urea was added into the reactor vessel, and the suspension was stirred continuously for 16 h. The deposition was followed by centrifugation of the catalyst suspension in 50 ml tubes. The centrifugation was conducted three times. Separated water was decanted away, and the tube was refilled with distilled water after the first and the second centrifugations. Posterior the following separation and washing, the solid was collected and moved to a rotary evaporator and dried at 60°C in a water bath under vacuum. Final drying was done in an oven at 120°C overnight. All catalysts were calcined in air flow by heating them from room temperature up to 300°C for 4 h. The synthesized monometallic catalysts will be named as AuAlDPU, AuCeDPU, AuAlCe_1DPU, AuAlCe_5DPU, and AuAlCe_{10}DPU.

The supported Ag nanoparticles were synthesized by DP with NaOH. The procedure was the same as the described for the gold synthesis by DP with urea, only that, instead of urea, NaOH was occupied, regulating solution's pH to 9. The synthesized monometallic catalysts will be named as AgCeDPNa, AgAlDPNa, AgAlCe_1DPNa, AgAlCe_3DPNa, AgAlCe_5DPNa, AgAl$Ce_{7.5}$DPNa, and AgAlCe_{10}DPNa. All the catalysts prepared are mentioned in **Table 1**.

2.4 Characterization of noble (Ru, Au, Ag) and base (Cu) metal nanoparticles supported

Figure 1 shows the adsorption isotherms of the synthesized materials of RuAlIH and RuAlCe_1IH. It was observed that both isotherms are of type IV, which were associated with capillary condensation in mesoporous catalysts, where the hysteresis loops indicated that the pores are well distributed.

For the catalysts of RuAlIH and RuAlCe_1IH (the other TPR analyzes the rest of the catalysts not shown), **Figure 2** which displayed a main peak of 36–52°C was

Catalyst abbreviation	Synthesis method						Target molecule		
	One step			Two step					
	Wet impregnation	Sol-gel	Wet impregnation with urea	Wet impregnation	Deposit-precipitation With urea	Deposit-precipitation With NaOH	MTBE	ETBE	TAME
Cu_5AlIH	X							x	x
$Cu_{10}AlIH$	X							x	x
$C_{15}AlIH$	X							x	x
Cu_5AlSG,		x						x	x
$Cu_{10}AlSG$		x						x	x
$Cu_{15}AlSG$		x						x	x
$Cu_{15}AlSGp$		x						x	x
Cu_5AlIHU			x					x	x
$Cu_{10}AlIHU$			x					x	x
$Cu_{15}AlIHU$			x					x	x
$Cu_{15}AlIHp$			x					x	x
RuAlIH				x			x		
$RuAlCe_1IH$				x			x		
$RuAlCe_3IH$				x			x		
$RuAlCe_5IH$				x			x		
$RuAlCe_{7.5}IH$				x			x		
$RuAlCe_{10}IH$				x			x		
AuAlDPU					x		x		x
AuCeDPU					x		x		x

Catalyst abbreviation	Synthesis method						Target molecule		
	One step			Two step					
	Wet impregnation	Sol-gel	Wet impregnation with urea	Wet impregnation	Deposit-precipitation With urea	Deposit-precipitation With NaOH	MTBE	ETBE	TAME
$AuAlCe_1DPU$					x		x		x
$AuAlCe_3DPU$					x		x		x
$AuAlCe_5DPU$					x		x		x
$AuAlCe_{10}DPU$					x		x		x
AgCeDPNa						x	x		
AgAlDPNa						x	x		
$AgAlCe_1DPNa$						x	x		
$AgAlCe_3DPNa$						x	x		
$AgAlCe_5DPNa$						x	x		
$AgAlCe_{7.5}DPNa$						x	x		
$AgAlCe_{10}DPNa$						x	x		

Table 1.
Lists the Cu, Ag, Au, and Ru catalysts supported on Al_2O_3, CeO_2, and Al_2O_3-CeO_2, tested in CWAO of fuel oxygenated (FO) with the operating conditions: T = 100°C, $P(O_2)$ = 10 bar, VLiq = 0.25 l, CFO = 1000 mg/l, CCat = 1 g/l, and ω = 1000 rpm.

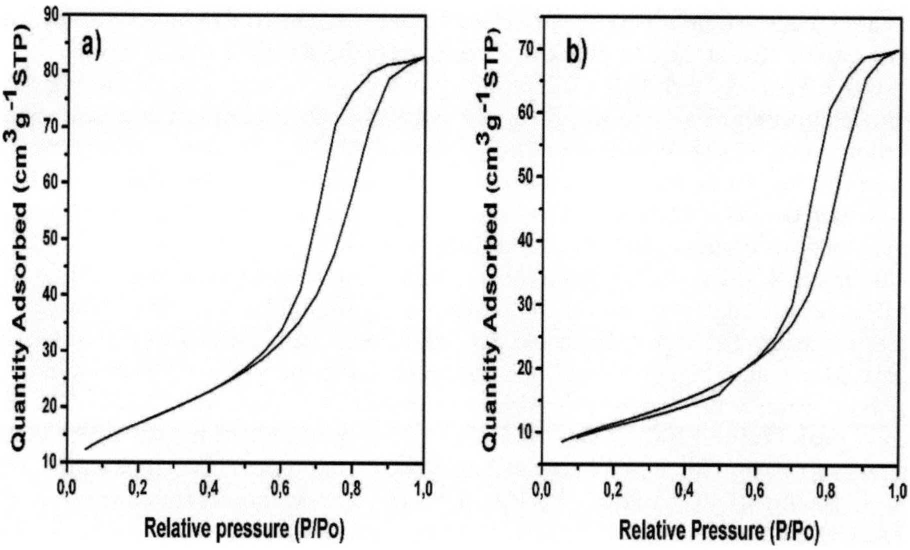

Figure 1.
Adsorption isotherms of (a) RuAlIH and (b) RuAlCe$_1$IH.

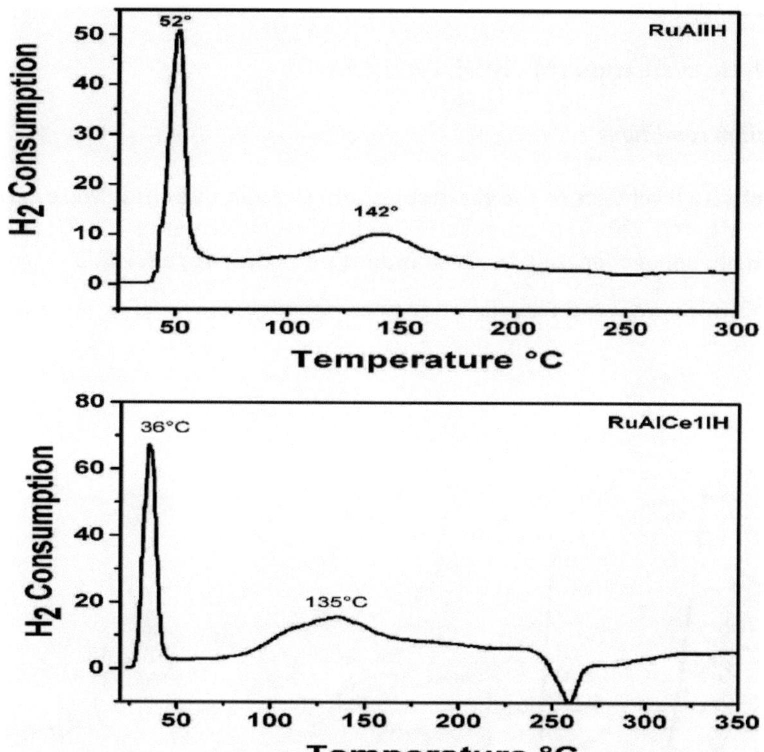

Figure 2.
H_2-TPR profiles of the catalysts RuAlIH and RuAlCe$_1$IH.

observed which indicates that the reduction is carried out in that first peak, and it was attributed to the oxidation change of Ru from +2 to 0 (RuO) since it was the species that was reduced first. The second signal observed at 135–142°C was attributed to ruthenium oxide (RuO_2), with an oxidation state of +4 which passes from +4 to +2 and subsequently to 0. On the other hand, the two peaks clearly observed in **Figure 2** indicated that the ions of Ru existed in two different states to be reduced with hydrogen, meaning that at the end of the reduction, only the states +2 and 0 remain.

Figure 3 corresponds to the diffraction patterns of the catalysts containing Au. It showed only signals corresponding to Al_2O_3 and CeO_2, and only a decrease of the alumina signal was observed when the content of Al_2O_3-CeO_2 increases by 10%. The corresponding Au signals were not shown in this diffractogram due to the weight % in which the catalysts were prepared, and in XRD only the metal was observed at concentrations higher than 2 and sometimes 3%.

In **Figure 4**, the H_2-TPR profiles of the Au-supported catalysts revealed that the first reduction peaks (around 50°C) appearing for all Au-supported catalysts corresponded to the highly dispersed Au peaks on the catalyst surface. This signal increased to values higher than 50°C in the case of the Au catalyst deposited in Ce which indicates a difference in the size of the particles (observed by TEM). The second peak (around 100°C) was attributed to a second oxidation state of Au that interacts with Ce. This signal increased with the Ce content. This can be supported since in the AuAlDPU catalyst, this signal did not appear; however, it appeared in the AuCeDPU catalyst.

Figure 5 shows the XRD for the copper catalysts prepared by wet impregnation method, in which γ-Al_2O_3 phase was seen as well as the intense signals that indicated the presence of CuO, and the boehmite, indicating that the metal was correctly dispersed in the three synthesized catalysts.

2.5. Catalytic evaluation (MTBE, ETBE, TAME)

2.5.1 Reaction conditions

The activity level tests of the catalysts synthesized in this study were carried out in a Parr batch 300 ml batch reactor, under the conditions of 100°C, 10 bar, and 1000 ppm of fuel oxygenated. In the standard procedure for a CWAO experiment,

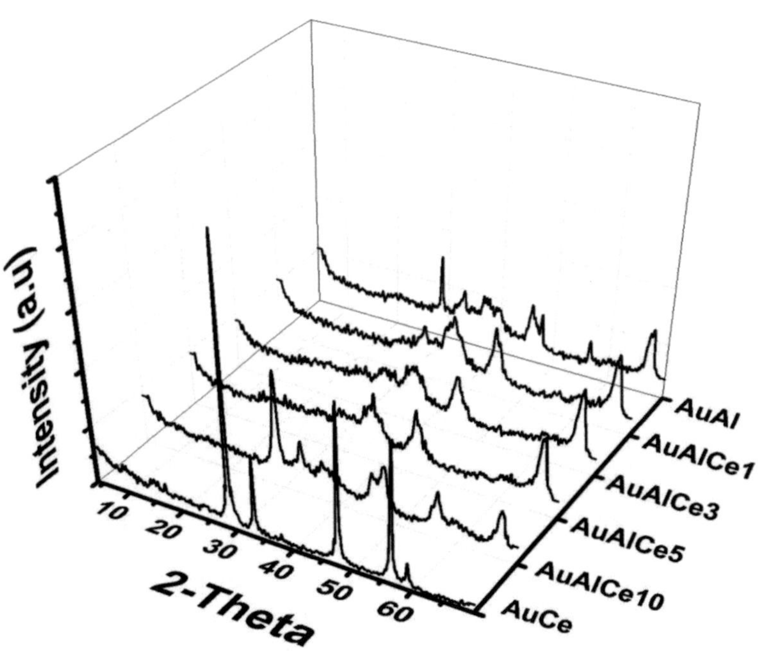

Figure 3.
X-ray diffraction patterns of Au-supported catalysts.

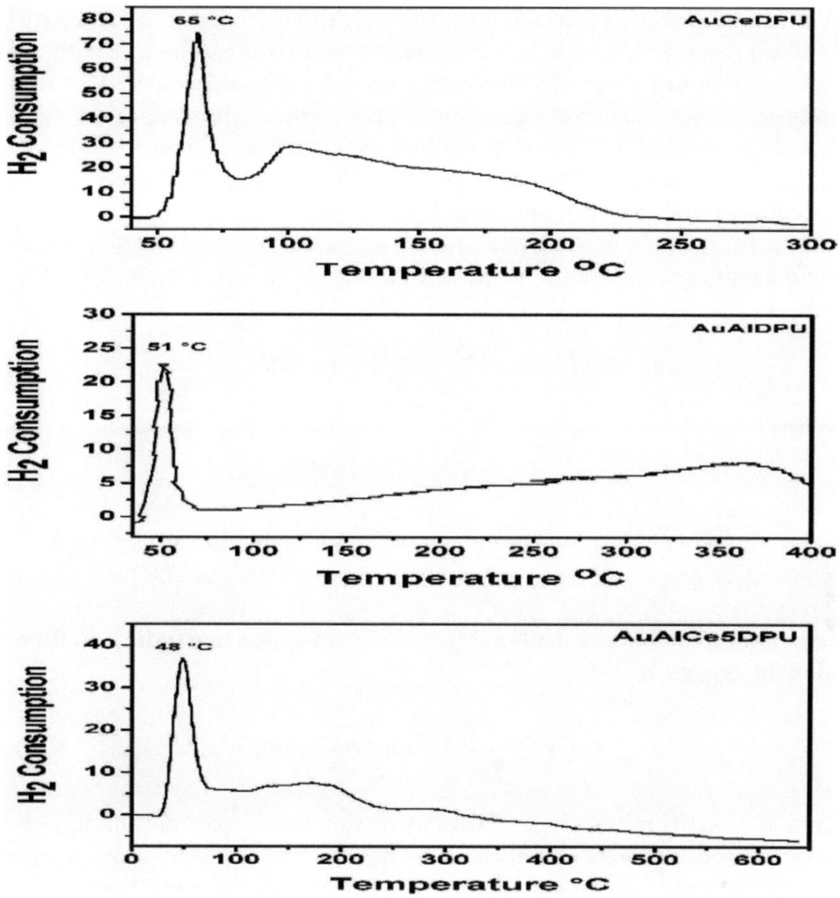

Figure 4.
H_2-TPR of Au-supported catalysts.

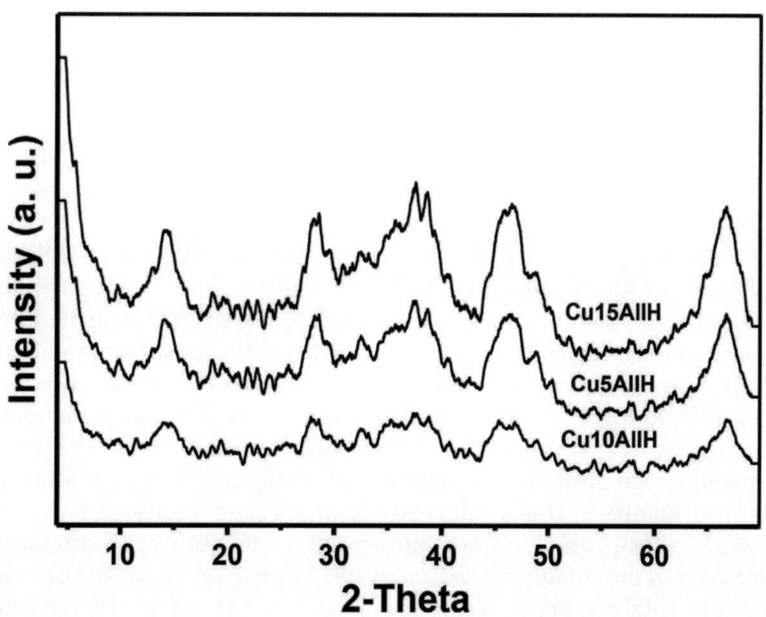

Figure 5.
X-ray diffraction patterns for the Cu-supported catalysts synthesized by wet impregnation in a single step.

250 ml of fuel oxygenated solution were poured, and 0.25 g of catalyst were placed in the 300 ml reactor. When the selected temperature was reached, stirring was started at a maximum speed of 1000 rpm. This time was taken as the zero reaction time and the reaction duration was 60 min. These conditions were the same for all synthesized materials. The liquid samples were periodically removed from the reactor, then filtered to remove any catalyst particles, and finally analyzed by gas chromatography and total organic carbon (TOC).

With the following equation, the conversion values for total organic carbon and FO were determined at different times with intervals of 30 min up to 180 min of reaction:

$$X_{TOC} = \frac{TOC^0 - TOC^{60}}{TOC^0} \times 100 \quad (1)$$

$$X_{FO} = \frac{C_0 - C_{60}}{C_0} \times 100\% \quad (2)$$

where TOC^0 is TOC at t = 0 (ppm), C_0 is FO concentration at t = 0 (ppm), C_{60} is FO concentration at t = 1 h of reaction (ppm), and TOC^{60} is TOC at t = 1 h of reaction (ppm).

The initial rate (ri) was calculated from FO conversion depending on time, using the following equation:

$$r_i = \left(\frac{\Delta_{FO}(\%)}{\Delta t\, m_{cat}}\right)([contaminant]_i) \quad (3)$$

where $\frac{\Delta_{FO}(\%)}{\Delta t}$ is the initial slope of the conversion curve, $[\text{contaminant}]_i$ = initial FO concentration, and m_{cat} = catalyst mass (g_{cat}/l).

So the selectivity was calculated according to the following equation:

$$S_{CO_2} = \frac{X_{TOC}}{X_{FO}} \times 100 \quad (4)$$

2.5.2. Degradation of MTBE by catalytic wet air oxidation over noble and base metals

This research studied the degradation of the fuel oxygenated MTBE through CWAO occupying Ru, Au, and Ag as the catalysts, which will be responsible for mineralizing the pollutant.

Figure 6 shows the results of the catalytic activity for CWAO of MTBE with the sets of catalysts Ru, Au, and Ag. The best activity for the set of ruthenium-supported catalysts was for the one named $RuAlCe_1IH$, since it presented a 68% conversion of MTBE and a 63% degradation of TOC; the most favorable results for this particular catalyst were attributed to a particle size of 9 nm measured by TEM, and to the contribution of ceria, due to the oxygen storage capacity phenomenon. Ceria is known for having the capacity to exchange oxygen, through its vacancies of oxygen, which promotes the increase of selectivity in all cases with the Ru-synthesized catalysts from a chlorinated salt, by the formation of a species of type $Ce^{4+} - O^{2-} - M^{+}$, contributing to improve the reducibility of ruthenium.

With respect to the Au-supported catalysts, $AuAlCe_5DPU$ was the one that stood out for its catalytic performance in comparison to the rest of its counterparts. $AuAlCe_5DPU$ reached a maximum of 73% MTBE conversion and a TOC degradation of 72%, indicating that this catalyst was efficient both to have a good conversion and to transform to CO_2. This behavior was explained by the fact of presenting the best distribution of particles on the surface of the catalyst, according to the TEM analysis, and confirmed by TPR. The largest amount of active particles for this catalyst was below 2 nm. It was observed that the other catalysts had a similar activity attributable to the distribution of particle sizes ranging between 2 and 10 nm. According to the performed analysis, well-dispersed Au nanoparticles and the oxidation state of Au play an important role in this type of oxide-reduction reactions. The excess of CeO_2 does not allow a good selectivity toward CO_2 since it interferes in the exchange of O_2 at the time of oxidation, giving an excess of oxygen that causes the metal particle to change its oxidation state on the surface of the catalyst and confirming the theory made by Imamura et al. [26] that a balance of metal particles in oxidized and reduced state is needed to obtain satisfactory results in terms of activity and selectivity; this theory is fulfilled in molecules that are strongly adsorbed on the surface of the catalyst such as acetic acid, and according to this study, this principle can be also applied with MTBE.

The results of the catalytic activity for CWAO of MTBE for the set of silver catalysts indicated that the best conversion obtained corresponds to the AgCeDPNa catalyst with 66%, due to the presence of CeO_2. In this case, we can only mention an effect of CeO_2 because the particle sizes obtained by HRTEM and TPD-CO revealed different distributions but did not significantly impact the catalytic activity. This effect has been explained by several researchers as the formation of a bridge M-O-Ce where M means silver metal. The AgCeDPNa catalyst was the one that has a better behavior toward mineralizing CO_2 due to the oxide-reducing properties of this support. However, it was observed that the catalyst with 5% of CeO_2 has a very close TOC value with respect to the AgCeDPNa catalyst. In this case, the effect between the particle size, the activity, and the selectivity toward CO_2 was not possible to distinguish because, as the HRTEM histograms showed, the range of sizes was

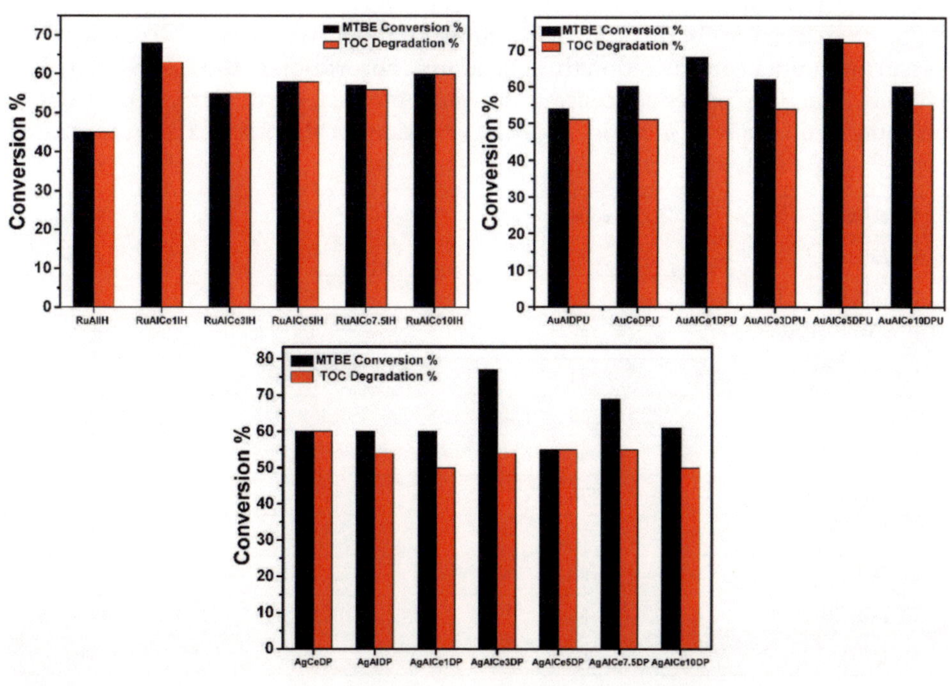

Figure 6.
MTBE conversion and TOC degradation, at 100°C and 10 oxygen bar over Ru-, Au-, and Ag-supported catalysts.

very broad in all cases; nonetheless, the effect of CeO_2 appeared in the activity and also in the selectivity. Imamura proposed a theory stating that a balance of metal particles in oxidized and reduced state is needed to obtain satisfactory results in terms of activity and selectivity, which is observed here by TPR of H_2 in the case of AgCeDPNa catalyst; Ag particles remain oxidized even with the passage of H_2 which makes them more selective toward CO_2. It is not possible in this case to account for the proportion of Ag^+/AgO particles because this can only be done by XPS, a very expensive technique and not available in this case; however, it can be concluded that, in terms of activity, the optimal catalyst is the one that only contains ceria.

2.5.3. Degradation of ETBE by catalytic wet air oxidation over noble and base metals

We also analyzed the degradation process of the ETBE molecule through CWAO using Cu synthesized by three different synthesis methods, with the main characteristic of being carried out in a single step. The aim of the application of these methods is to avoid the leaching of the metal, which has occurred in other previous experiments by different investigators, after having passed a certain time of the reaction.

The analysis results of the ETBE-treated solutions by CWAO are presented in **Table 2**; for each of the three synthesis methods of the Cu catalysts, Cu_{10}AlSG and Cu_{10}AlIHU catalysts were more active, obtaining 88 and 89% ETBE conversion, respectively, after 1 h of oxidation. But the highest values for TOC degradation were obtained with the catalysts prepared by sol-gel method, as Cu_5AlSG reached 84% and Cu_{10}AlSG 89%. This last result confirmed that Cu catalysts synthesized by sol-gel are more effective catalysts for the mineralization process, which allows to degrade the organic matter, coming from the contaminant present, by almost 90% until obtaining CO_2 in the treated solutions. In addition, we can affirm that the optimum percentage of Cu was 10%, as well as the commercial catalyst used for this type of reactions and reported in the literature.

Another discovery in this study with copper nanoparticles was realized by measuring copper concentration through atomic absorption in the ETBE-treated solutions, since no copper concentrations were obtained, particularly in the catalysts prepared by sol-gel method, opposite situation for the other catalysts prepared

Catalyst	ETBE conversion %	% TOC	SCO_2
Cu_5AlIH	76	81	100
Cu_{10}AlIH	66	82	100
C_{15}AlIH	72	80	100
Cu_5AlSG	82	84	100
Cu_{10}AlSG	88	89	100
Cu_{15}AlSG	80	78	97
Cu_{15}AlSGp	78	74	95
Cu_5AlIHU	71	59	83
Cu_{10}AlIHU	89	79	89
Cu_{15}AlIHU	86	81	94
Cu_{15}AlIHp	80	69	86

Table 2.
ETBE conversion, TOC and SCO_2 at 100°C and 10 bar of pressure during 1 h of reaction with one of $[ETBE]_0$ = 1000 ppm.

by wet impregnation and wet impregnation with urea. Therefore, we can encapsulate the copper in the alumina, by sol-gel method, thus avoiding contamination by the metal, leaching, and, in turn, obtaining an improvement in the catalytic performance of the metal.

2.5.4. Degradation of TAME by catalytic wet air oxidation over noble and base metals

Another series of experiments was conducted under the same conditions described below over Cu synthetized in three different methods and Au-supported catalyst but in this experiments with TAME as a target molecule by CWAO.

Table 3 presents the analysis results of the treated solutions of TAME by CWAO; for each of the three synthesis methods of the Cu catalysts, the catalysts of $Cu_{15}AlSG$ and $Cu_{10}AlIHU$ were more active, obtaining 78% of TAME conversion, for both catalysts after 1 h of reaction. But the highest values for TOC degradation were obtained with the catalysts prepared by sol-gel method; $Cu_{15}AlSG$ reached 78% and $Cu_{10}AlIHU$ 75%. These results sustained that the Cu catalysts synthesized by sol-gel were more effective catalysts for TAME mineralization process, which allows them to degrade the organic matter, coming from the existing contaminant, by almost 80% in the treated solutions.

The results of the catalytic activity for TAME CWAO in Au-supported catalysts are shown in **Figure 7**. It was observed that the best activity happened with the catalyst at $AuAlCe_{10}DPU$ with 80% conversion of TAME, although all the remaining catalysts showed good activity except for the catalyst with $AuAlCe_1DPU$; this could be explained by the fact that this molecule may not be very sensitive to particle size due to its structure.

Figure 7 also shows the abatement of TOC of the supported Au catalysts; as can be seen the best carbon transformation toward CO_2 was obtained for the $AuAlCe_3DPU$ catalyst, with a 77% conversion, and for the AuCeDPU catalyst with 80%, although all the remaining catalysts showed a remarkable performance, without exceeding these, except for $AuAlCe_1DPU$.

Table 4 shows the selectivity to CO_2, and we can say that the supported Au catalysts containing Ce 5 and 10% were the least selective to CO_2. This is because there is

Catalyst	TAME conversion %	TOC	SCO_2
Cu_5AlIH	72	68	93
$Cu_{10}AlIH$	68	61	89
$C_{15}AlIH$	77	75	98
Cu_5AlSG	73	68	94
$Cu_{10}AlSG$	72	68	95
$Cu_{15}AlSG$	78	78	100
$Cu_{15}AlSGp$	70	41	59
Cu_5AlIHU	49	64	100
$Cu_{10}AlIHU$	78	75	95
$Cu_{15}AlIHU$	66	62	95
$Cu_{15}AlIHp$	69	34	50

Table 3.
TAME conversion, TOC degradation, and SCO_2 at 100°C and 10 bar of pressure during 1 h of reaction with one of $[TAME]_0$ = 1000 ppm.

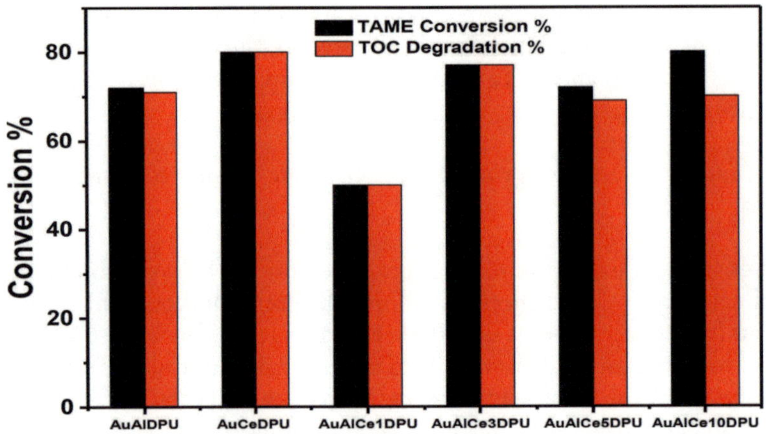

Figure 7.
TAME conversion and TOC degradation % at 100°C and 10 bar over Au-supported catalysts.

Catalysts	SCO_2
AuAlDPU	99
AuCeDPU	100
$AuAlCe_1DPU$	100
$AuAlCe_3DPU$	100
$AuAlCe_5DPU$	96
$AuAlCe_{10}DPU$	88

Table 4.
TAME selectivity at 100°C over Au-supported catalysts.

a poisoning by CeO_2 that affects the selectivity when it is in excess due to the interaction of M-Ce-O; this case shows that the optimal percentage of CeO_2 for the TAME is 3%. It is important to note that the $AuAlCe_3DPU$ catalyst is equally active and selective to the AuCeDPU catalyst, so the alumina-ceria support with low concentrations of CeO_2 is presented as an alternative for the wet oxidation process of TAME.

3. Conclusions

This study concludes that the catalytic activity for MTBE oxidation of catalysts Ru, Au, and Ag, supported on Al_2O_3, CeO_2, and Al_2O_3-CeO_2, synthesized by wet impregnation methods and DP with NaOH and urea, in two steps, is classified as follows:

$$AuAlCe_5\,DPU > RuAlCe_1\,IH > AgCeDPNa \tag{5}$$

In addition, the catalytic activity for the oxidation of target molecule ETBE on Cu catalysts supported on Al_2O_3, synthesized by three different methods in a single step, is classified as follows:

$$Cu_{10}\,AlSG > Cu_{10}\,AlIH > Cu_{15}\,AlIHU \tag{6}$$

The catalytic activity for TAME oxidation using Copper catalysts supported on Al_2O_3, synthesized by three different methods in a single step and Au supported on

Al_2O_3, CeO_2 and Al_2O_3-CeO_2 synthesized by DP with urea in two steps, is classified as follows:

$$AuCeDPU > Cu_{15}AlSG > AuAlCe_3DPU > Cu_{10}AlIHU = Cu_{15}AlIH \quad (7)$$

Acknowledgements

Authors thank the National Council for Science and Technology (CONACYT) for financing the project 132648; thanks to the Universidad Juárez Autónoma de Tabasco for PFCE-DACB project and PRODEP program.

Conflict of interest

Authors declare no conflicts of interest.

Author details

Zenaida Guerra Que, José Gilberto Torres Torres*, Ignacio Cuauhtémoc López, Juan C. Arévalo Pérez, Adrian Cervantes Uribe, Hermicenda Pérez Vidal, Alejandra E. Espinosa de los Monteros Reyna, José G. Pacheco Sosa, María A. Lunagómez Rocha and Cecilia Sánchez Trinidad
Laboratory of Catalytic Nanomaterials Applied to the Development of Energy Sources and Environmental Remediation, Applied Science and Technology Research Center of Tabasco (CICTAT), DACB, Juarez Autonomous University of Tabasco, Tabasco, México

*Address all correspondence to: gilberto.torres@ujat.mx

References

[1] Awad OI, Mamat R, Ibrahim TK, Hammid AT, Yusri IM, Hamidi MA, et al. Overview of the oxygenated fuels in spark ignition engine: Environmental and performance. Renewable and Sustainable Energy Reviews. 2018;**91**:394-408. DOI: 10.1016/j.rser.2018.03.107

[2] McGregor D. Fuel Oxygenates. Encyclopedia of Toxicology. 3rd ed. Vol 2; USA: Academic Press, Elsevier Inc; 2014. pp. 671-681. ISBN: 9780123864543. DOI: 10.1016/b978-0-12-386454-3.00025-7

[3] Cuauhtémoc I, Del Angel G, Torres G, Navarrete J, Angeles-Chavez C, Padilla JM. Synthesis and characterization of Rh/Al_2O_3-CeO_2 catalysts: Effect of the Ce^{4+}/Ce^{3+} ratio on the MTBE removal. Journal of Ceramic Processing Research. 2009;**10**:512-520

[4] Bergendahl JA, Thies TP. Fenton's oxidation of MTBE with zero-valent iron. Water Research. 2004;**38**:327-334. DOI: 10.1016/j.watres.2003.10.003

[5] Mehrjouei M, Müller S, Möller D. Decomposition kinetics of MTBE, ETBE and, TAEE in water and wastewater using catalytic and photocatalytic ozonation. Journal of Molecular Catalysis A: Chemical. 2014;**386**:61-68. DOI: 10.1016/j.molcata.2014.02.014

[6] Bartling J, Esperschütz J, Wilke BM, Schloter M. ETBE (ethyl tert butyl ether) and TAME (tert amyl methyl ether) affect microbial community structure and function in soils. Journal of Hazardous Materials. 2011;**187**: 488-494. DOI: 10.1016/j.jhazmat.2011.01.058

[7] Cuauhtémoc I, Del Angel G, Torres G, Bertin V. Catalytic wet air oxidation of gasoline oxygenates using Rh/γ-Al_2O_3 and Rh/γ-Al_2O_3-CeO_2 catalysts. Catalysis Today. 2008;**133-135**:588-593. DOI: 10.1016/j.cattod.2008.02.006

[8] Cuauhtémoc I, Del AG, Torres G, Angeles-chavez C, Navarrete J, Padilla JM. Enhancement of catalytic wet air oxidation of tert-amyl methyl ether by the addition of Sn and CeO_2 to Rh/Al_2O_3 catalysts. Catalysis Today. 2011;**166**: 180-187. DOI: 10.1016/j.cattod.2010.11.100

[9] Kasprzyk-Hordern B, Andrzejewski P, Dabrowska A, Czaczyk K, Nawrocki J. MTBE, DIPE, ETBE and TAME degradation in water using perfluorinated phases as catalysts for ozonation process. Applied Catalysis B: Environmental. 2004;**51**:51-66. DOI: 10.1016/j.apcatb.2004.02.004

[10] Sun Y, Zhang Y, Quan X. Treatment of petroleum refinery wastewater by microwave-assisted catalytic wet air oxidation under low temperature and low pressure. Separation and Purification Technology. 2008;**62**: 565-570. DOI: 10.1016/j.seppur.2008.02.027

[11] Fathy NA, El-Khouly SM, Hassan NA, Awad RMS. Free- and Ni-doped carbon xerogels catalysts for wet peroxide oxidation of methyl orange. Journal of Water Process Engineering. 2017;**16**:21-27. DOI: 10.1016/j.jwpe.2016.11.005

[12] Redel-Macías MD, Pinzi S, Leiva-Candia DE, López I, Dorado MP. Ternary blends of diesel fuel oxygenated with ethanol and castor oil for diesel engines. Energy Procedia. 2017;**142**:855-860. DOI: 10.1016/j.egypro.2017.12.137

[13] Abasaeed AE, Al-Fatesh AS, Naeem MA, Ibrahim AA, Fakeeha AH.

Catalytic performance of CeO_2 and ZrO_2 supported Co catalysts for hydrogen production via dry reforming of methane. International Journal of Hydrogen Energy. 2015;**40**(21):6818-6826. DOI: 10.1016/j.ijhydene.2015.03.152

[14] Cozzarelli IM, Mckelvie JR, Baehr AL. Volatile Hydrocarbons and Fuel Oxygenates. Treatise on Geochemistry. 2nd ed. Vol 11, 12. Oxford: Elsevier; 2014. pp. 439-480. DOI: 10.1016/b978-0-08-095975-7.00912-8

[15] Levchuk I, Bhatnagar A, Sillanpää M. Overview of technologies for removal of methyl tert-butyl ether (MTBE) from water. Science of the Total Environment. 2014;**476-477**:415-433. DOI: 10.1016/j.scitotenv.2014.01.037

[16] Van Afferden M, Rahman KZ, Mosig P, et al. Remediation of groundwater contaminated with MTBE and benzene: The potential of vertical-flow soil filter systems. Water Research. 2011;**45**:5063-5074. DOI: 10.1016/j.watres.2011.07.010

[17] Massa A, Hernández S, Ansaloni S, Castellino M, Russo N, Fino D. Enhanced electrochemical oxidation of phenol over manganese oxides under mild wet air oxidation conditions. Electrochimica Acta. 2018;**273**:53-62. DOI: 10.1016/j.electacta.2018.03.178

[18] Van Afferden M, Rahman KZ, Mosig P, De Biase C, Thullner M, Oswald SE, et al. Remediation of groundwater contaminated with MTBE and benzene: The potential of vertical-flow soil filter systems. Water Research. 2004;**38**(16):5063-5074. DOI: 10.1016/j.watres.2011.07.010

[19] Sutherland J, Adams C, Kekobad J. Treatment of MTBE by air stripping, carbon adsorption, and advanced oxidation: Technical and economic comparison for five groundwaters. Water Research. 2004;**38**:193-205. DOI: 10.1016/j.watres.2003.09.008

[20] Tsoncheva T, Ivanova R, Henych J, Dimitrov M, Kormunda M, Kovacheva D, et al. Effect of preparation procedure on the formation of nanostructured ceria—Zirconia mixed oxide catalysts for ethyl acetate oxidation: Homogeneous precipitation with urea vs template-assisted hydrothermal synthesis. Applied Catalysis A: General. 2015;**502**:418-432. DOI: 10.1016/j.apcata.2015.05.034

[21] Roy B, Martinez U, Loganathan K, Datye AK, Leclerc CA. Effect of preparation methods on the performance of Ni/Al_2O_3 catalysts for aqueous-phase reforming of ethanol: Part I-catalytic activity. International Journal of Hydrogen Energy. 2012;**37**:8143-8153. DOI: 10.1016/j.ijhydene.2012.02.056

[22] Benito P, Gregori M, Andreoli S, Fornasari G, Millefanti S, Ospitali F, et al. Role of the preparation method on properties of Pd/Cu-MCM-41 hydrodechlorinating catalysts. Catalysis Today. 2014;**235**:134-143. DOI: 10.1016/j.cattod.2014.01.034

[23] Manzoli M, Menegazzo F, Signoretto M, Cruciani G, Pinna F. Effects of synthetic parameters on the catalytic performance of Au/CeO_2 for furfural oxidative esterification. Journal of Catalysis. 2015;**330**:465-473. DOI: 10.1016/j.jcat.2015.07.030

[24] Levec J, Pintar A. Catalytic wet-air oxidation processes: A review. Catalysis Today. 2007;**124**(3-4):172-184. DOI: 10.1016/j.cattod.2007.03.035

[25] Kim SK, Kim KH, Ihm SK. The characteristics of wet air oxidation of phenol over $CuOx/Al_2O_3$ catalysts:

Effect of copper loading. Chemosphere. 2007;**68**:287-292. DOI: 10.1016/j.chemosphere.2006.12.080

[26] Hosokawa S, Kanai H, Utani K, Taniguchi YI, Saito Y, Imamura S. State of Ru on CeO_2 and its catalytic activity in the wet oxidation of acetic acid. Applied Catalysis B: Environmental. 2003;**45**:181-187. DOI: 10.1016/S0926-3373(03)00129-2

Chapter 4

Removal of Cr(VI) from Waters by Multi-Walled Carbon Nanotubes: Optimization and Kinetic Investigations

Francisco J. Alguacil and Félix A. Lopez

Abstract

The adsorption of chromium(VI) from aqueous solutions onto multi-walled carbon nanotubes (MWCNTs) has been investigated under various experimental conditions of initial metal concentration, agitation speed, aqueous pH, temperature and adsorbent dosage to assess the equilibrium and kinetic parameters. It was found that the kinetic data were fitted with the pseudo-first- and pseudo-second-order models, whereas the chromium(VI) adsorption data were fitted with the Langmuir and Freundlich equilibrium models to give the characteristic parameters of each model. According with the evaluation, both isotherm models are useful to represent the measured adsorption data. The adsorption of chromium(VI) is also dependent on the temperature, and the corresponding thermodynamic parameters including $\Delta H°$, $\Delta G°$ and $\Delta S°$ were estimated from the experimental data, indicating the exothermic and non-spontaneous nature of the metal adsorption onto the MWCNTs. Chromium(VI) desorption was investigated by the use of aqueous hydrazine sulfate solutions.

Keywords: chromium(VI), multi-walled carbon nanotubes, adsorption, desorption, kinetics

1. Introduction

Despite its toxic character, chromium(VI) is widely used in various industries, being its recovery from the corresponding liquid effluents a primary target before their discharge to natural waters. Several technologies have found application to remove and/or recover chromium(VI) from these process wastes: pseudo-emulsion strip dispersión pertraction [1, 2], adsorption onto activated carbons [3], liquid-liquid extraction [4], biomass adsorption [5, 6], adsorption onto natural zeolites [7], adsorption onto phosphates [8], ion exchange [9] and electro-assisted and photo-assisted technologies [10]. Among them, adsorption onto carbon nanotube (CNT) technology could be competitive when the metal is present at low concentrations in the aqueous solution. Various carbon nanotubes configurations can be found, being the most commonly used the single-walled carbon nanotubes (SCNTs) and multi-walled carbon nanotubes (MWCNTs) configurations, together with functionalized multi-walled carbon nanotubes. In either configuration, and after the metal adsorption, a subsequent operation or elution is needed in order to recover the metal to a solution where it is concentrated and purified, and thus it can be conveniently recovered or even recycled to the original industrial process.

A number of examples using these CNTs for chromium(VI) recovery from aqueous solutions can be found in the literature. Naghizadeh [11] investigated the adsorption efficiency of activated carbon and multi-walled carbon nanotubes respect to cadmium(II) and chromium(VI) in the 3–12 pH range. Whereas both adsorbents presented high metal adsorption capacities over the whole pH range investigated, the experimental results indicated that MWCNTs had a greater potential for the removal of chromium(VI) and cadmium(II) from aqueous solutions than activated carbon. In an investigation by [12], oxidized (–COOH) multi-walled carbon nanotubes were used to remove Cr(VI) (hazardous element) and Au(III) (valuable element), from aqueous solutions. Experiments were performed in order to investigate the influence of different variables on the adsorption kinetics, i.e., the stirring speed (250–2000 min^{-1}) and adsorbent dosage (0.25–1.5 g/L) in the case of chromium(VI) as well as temperature (20–60°C) and HCl (0.1–10 M) concentration in the case of gold(III). The performance of these carbon nanotubes was excellent in the removal of both elements, presented as the anions $HCrO_4^-$ and $AuCl_4^-$, from the aqueous solutions.

Anastopoulos et al. [13] reviewed the removal of chromium(III) and (VI) from aqueous solutions by carbon nanotubes. In both cases, the pH of the solution seemed to control the adsorption process, with a maximum adsorption of Cr(VI) occurring at pH 1–4 (in the case of Cr(III), the above occurs at pH values of 5–8). Furthermore, it is stated that most of the investigations are reported using non-real wastewater, conditions that very often are not repeated in real wastewaters. Xing et al. [14] presented a novel remediation protocol for Cr(VI) featured with high-capacity adsorption and electrochemical regeneration of the adsorbent. In their study, MWCNTs modified carbon cloth (CC) is used as a useful carrier for electrodepositing polypyrrole (PPy) film and the resultant nanocomposite CC-MWCNTs-PPy is used as an adsorbent with high adsorption capacity and stability. CC-MWCNTs-PPy is electrically regenerated to reduce secondary wastes.

In the present work, results obtained for the adsorption of chromium(VI) using multi-walled carbon nanotubes are presented. Several variables that could affect the adsorption process, such as the stirring speed of the aqueous solution, metal concentration and adsorbent dosage, temperature, etc., are investigated. Several equilibrium, kinetics and thermodynamic parameters are also reported. The desorption of the Cr(VI)-loaded MWCNTs is accomplished using aqueous solutions of hydrazine sulfate.

2. Methods

2.1 Reactives and experimental procedure

The multi-walled carbon nanotubes were obtained from Fluka and were used without further purification; the main characteristics of the adsorbent are given in **Table 1**, with further characteristics (i.e., Raman data) of them published elsewhere [15]. The characteristics of other adsorbent-ion exchangers used in this investigation were described elsewhere: Dowex 1x8 resin [16], oxidized MWCNTs [12] and activated carbon [17].

Stock Cr(VI) solutions were prepared by dissolving $K_2Cr_2O_7$ (Merck) in distilled water. All other chemicals were of AR grade.

Type	Multi-walled
Melting range	3652–3697°C
Density	2.1 g/mL
Appearance	Dust
Purity	≥98% carbon basis
Dimensions	10 ± 1 nm external diameter
Maximum adsorption	4.5 ± 0.5 nm internal diameter
BET	3–6 μm (length)
	1295 cm^3/g
	263 m^2/g

Table 1.
Characteristics of the multi-walled carbon nanotubes.

Metal adsorption (and elution) studies were carried out in a glass reactor provided for mechanical shaking. Metal adsorption (or elution) was determined by monitoring concentration by AAS in the aqueous solution as a function of time, whereas the metal concentration in the adsorbent was calculated by mass balance.

2.2 Modeling of kinetic adsorption

2.2.1 Pseudo-first-order model

The pseudo-first-order equation [18] used in this work can be expressed accordingly with the next equation:

$$\ln([Cr]_{c.e} - [Cr]_{c,t}) = \ln[Cr]_{c,e} - k_1 t \quad (1)$$

where $[Cr]_{c,t}$ and $[Cr]_{c,e}$ are the chromium concentrations in the nanotubes at equilibrium and at an elapsed time, respectively, t is the time and k_1 is the constant related to this model.

2.2.2 Pseudo-second-order model

In this model, the equation used is

$$\frac{t}{[Cr]_{c,t}} = \frac{1}{k_2 [Cr]^2_{c,e}} + \frac{t}{[Cr]_{c,e}} \quad (2)$$

In this case, k_2 is the constant related to this model.

2.3 Modeling the rate law

Three possible adsorption mechanisms had been evaluated if the adsorption of chromium(VI) into the MWCNTs must be considered as a liquid-solid phase reaction which includes diffusion of chromium species from the aqueous phase to the adsorbent surface, the diffusion of ions within the nanotubes and the chemical reaction between ions and any functional group in the carbon nanotubes [19]. The rate equations for the above three cases are:

i. film-diffusion controlled process, in which the rate equation is

$$\ln(1 - F) = -kt \tag{3}$$

ii. particle-diffusion controlled process, with the equation as

$$\ln(1 - F^2) = -kt \tag{4}$$

iii. Shrinking core model

$$3 - 3(1 - F)^{2/3} - 2F = kt \tag{5}$$

In all the above equations, F is the fractional approach to equilibrium, which is defined as

$$F = \frac{[Cr]_{c,t}}{[Cr]_{c,e}} \tag{6}$$

whereas k is the corresponding rate constant.

2.4 Modeling of adsorption isotherms

Both the Langmuir and Freundlich approaches had been used to model the experimental data, being both widely used in the modeling of adsorption or ion exchange processes [20].

The Langmuir model is valid for monolayer adsorption onto a surface containing a limited number of identical sites. The equation in its linear form describing this model is

$$\frac{1}{[Cr]_{c,e}} = \frac{1}{[Cr]_{c,m}} + \frac{1}{b[Cr]_{c,m}}\frac{1}{[Cr]_{s,e}} \tag{7}$$

where b is a constant related to the model, $[Cr]_{c,m}$ is the maximum metal uptake in the carbon nanotubes and $[Cr]_{s,e}$ is the equilibrium chromium(VI) concentration in the solution.

The Freundlich model is an empirical expression describing adsorption onto heterogeneous surfaces, having the adsorbent surface sites with a variation of binding energies. In this case, the equation also in its linear form is

$$\ln[Cr]_{c,e} = \ln k_f + \frac{1}{n}\ln[Cr]_{c,e} \tag{8}$$

where k_f and n are parameters related to the Freundlich model.

3. Results and discussion

3.1 Effect of stirring speed

Adsorption of chromium(VI) from aqueous solution to MWCNTs as a function of the stirring speed at pH 1 ± 0.1 is shown in **Figure 1**. The adsorption of chromium(VI) increases with increasing stirring speed, though from 1000 min^{-1} no significant

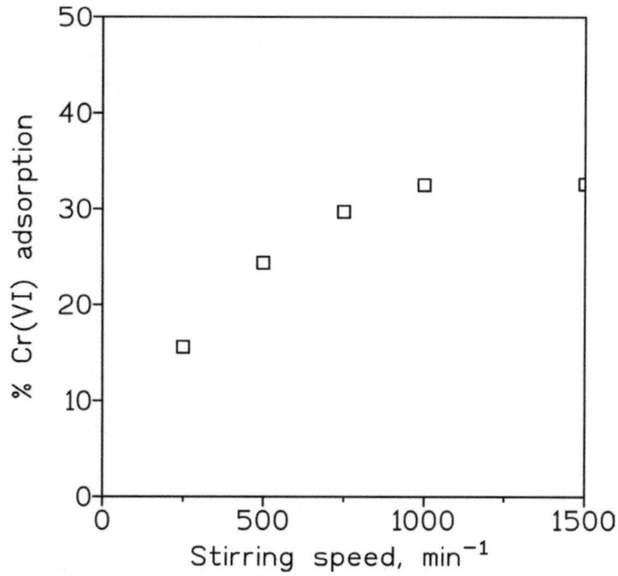

Figure 1.
Influence of stirring speed on the percentage of chromium(VI) adsorption at the equilibrium. Aqueous solution: 0.01 g/L Cr(VI). MWCNTs dosage: 1 g/L. Temperature: 20°C. Time: 2 h.

changes are encountered in metal adsorption specially at the longer contact times. These results shown that from 1000 min^{-1}, the thickness of the aqueous diffusion layer and the aqueous resistance to mass transfer were minimized, and the diffusion contribution of the aqueous species to the adsorption process is assumed to be constant.

3.2 Effect of temperature

The relationship between chromium(VI) adsorption and the temperature is also studied using aqueous solutions containing 0.01 g/L Cr(VI) at pH 4 ± 0.1 and adsorbent dosage of 1 g/L. **Table 2** shows the variation of log D_{Cr} vs. T, over the range of temperatures used, where D (the distribution coefficient) was calculated as

$$D = \frac{[Cr]_{c,e}}{[Cr]_{s,e}} \tag{9}$$

where $[Cr]_{c,e}$ and $[Cr]_{s,e}$ being the chromium concentrations in the nanotubes and in the aqueous solution at equilibrium. There is a decrease of chromium adsorption with the increase of temperature. One explanation of these results is to consider the nature of the species with the temperature as predicted by the Bjerrum equation. Accordingly, the estimated change of enthalpy is −14 kJ/mol, and the adsorption process is therefore exothermic.

Temperature	% adsorption	D (L/g)	Log D
20°C	44	0.79	−0.10
40°C	36	0.56	−0.25
60°C	28	0.39	−0.41

Stirring speed: 1000 min^{-1}. Time: 2 h.

Table 2.
Influence of temperature on chromium(VI) adsorption onto the MWCNTs.

The kinetic adsorption data were simulated with the two models shown in Eqs. (1) and (2), representing the pseudo-first- and pseudo-second-order models, respectively. The results are listed in **Table 3**. From the values of r2, the kinetic adsorption of chromium(VI) at the temperatures of 20 and 60°C can be fitted by the pseudo-second-order model.

3.3 Effect of pH

The pH of the aqueous solution may be one of the most decisive parameters controlling the adsorption process. The influence of pH on the adsorption of chromium(VI) is investigated at pH values ranging from 1 to 13. **Figure 2** shows that the maximum adsorption of the metal occurs at pH 4, and decreases either at more acidic and at alkaline pH values, these mean that $HCrO_4^-$ species (which is predominant at this range of initial chromium(VI) concentration and pH values below 6) is adsorbed onto the MWCNTs better than CrO_4^{2-} species, which is predominant at alkaline pH values. Furthermore, **Table 4** presented data about the adsorption of 0.005 g/L chromium(VI) at pH values of 1 and 4; it is also observed how the percentage of metal adsorption is greatly dependent on the pH of the aqueous solution, decreasing as the pH shifts to more acidic values.

3.4 Effect of carbon nanotubes dosage

It is apparent that the amount of adsorbent used in the removal of a given solute from aqueous solutions is critical for the practical application of such system. Thus, adsorption of chromium(VI) as a function of MWCNT dosages at pH 4 ± 0.1 is shown in **Figure 3**. The percentage of metal adsorption increases with the MWCNT dosage increasing, i.e. near 90% chromium(VI) is adsorbed at the adsorbent dosage of 10 g/L and this value down until 44% when the adsorbent dose is 1 g/L. These results are consistent with the fact that the increase of the adsorbent dosage results in the increase of the active sites in which the metal can be adsorbed, thus increasing the percentage of metal adsorbed or eliminated from the aqueous solution.

The data of the amount of chromium(VI) adsorbed on the MWCNTs (mg/g) and the metal concentration remaining in solution (mg/L) are fitted to the Langmuir and Freundlich models represented by Eqs. (7) and (8), respectively. The relative parameters obtained from the fit are listed in **Table 5**. The experimental data are well described by both models, indicating that the chromium(VI) uptake onto the MWCNTs is homogeneous and multilayer in nature. However, a singular fact of the Langmuir model can be described by the dimensionless separation factor, defined as

$$R = \frac{1}{1 + b\,[Cr]_{s,0}} \tag{10}$$

		20°C	60°C
Pseudo-first order	k_1 (min^{-1})	0.057	0.056
	r^2	0.9493	0.9856
Pseudo-second order	k_2 (g/min mg)	0.16	0.034
	r^2	0.9992	0.9983

Table 3.
Constants for the kinetic adsorption of chromium(VI) to MWCNTs using different adsorption models.

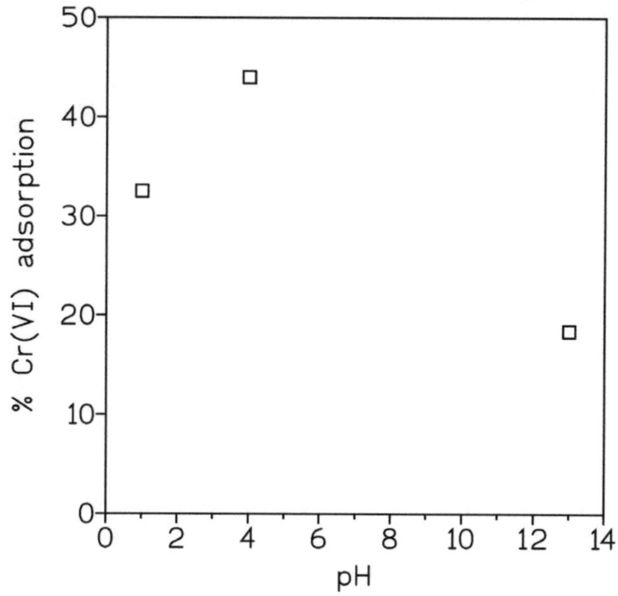

Figure 2.
Influence of the pH on chromium(VI) adsorption. Experimental conditions as in **Figure 1.**

pH ± 0.1	% adsorption
1	19.5
2	80.7
3	91.1
4	99.5
MWCNTs dosage: 10 g/L. Temperature: 20°C. Time: 2 h.	

Table 4.
Influence of pH on chromium(VI) adsorption onto the MWCNTs.

where $[Cr]_{s,0}$ is the initial metal concentration in the solution and b is the Langmuir constant. The value of R indicates if the adsorption is unfavorable (R > 1), linear (R = 1), favorable (0 < R < 1) or irreversible (R = 0). The value of R in this investigation was found to be 0.82, indicating that the adsorption of chromium(VI) is favorable.

3.5 Effect of metal concentration

The various adsorptions of chromium(VI) on MWCNTs as a function of initial metal concentration at pH 4 ± 0.1 are shown in **Figure 4**. The adsorption percentage of chromium(VI) decreases with initial metal concentration increasing.

The rate law governing the metal adsorption was investigated using the three models depicted in Eqs. (3)–(5), and the results from these fits were summarized in **Table 6**. It can be seen that within the particle-diffusion controlled model, the chromium(VI) adsorption onto the MWCNTs was better explained.

3.6 Comparison with other adsorbent-anion exchangers

The adsorption capacity, in terms of percentage of adsorption, found in this investigation was compared with the results obtained using other potential adsorbent-anion exchangers for Cr(VI). The results obtained from this set of

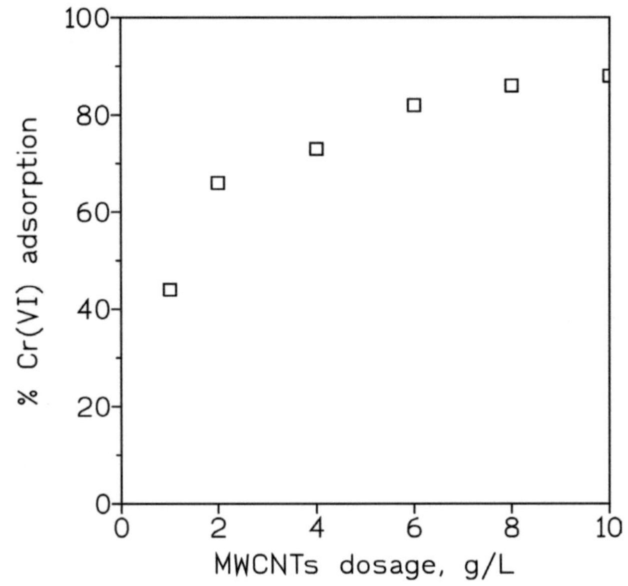

Figure 3.
Influence of MWCNTs dosage on chromium(VI) adsorption. Aqueous solution: 0.01 g/L Cr(VI) at pH 4. Temperature: 20°C. Time: 2 h.

	b (L/mg)	$[Cr]_{c,m}$ (mg/g)	r^2	ln k_f	1/n	r^2
Langmuir	0.021	37	0.9950			
Freundlich				−0.28	0.94	0.9952

Table 5.
Langmuir and Freundlich constants.

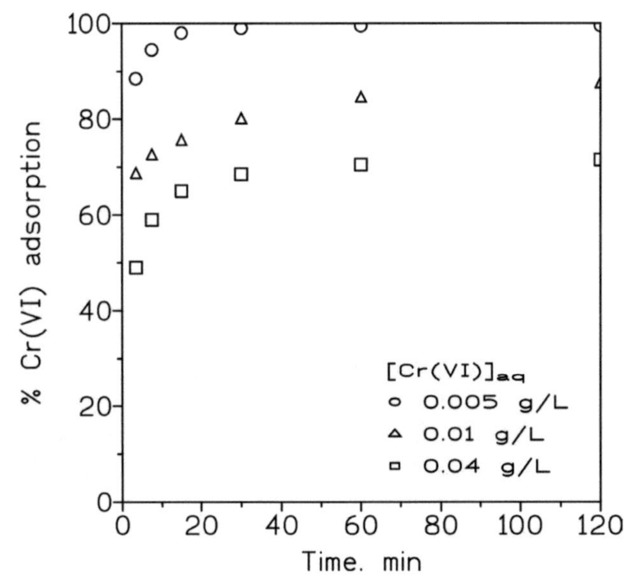

Figure 4.
Influence of initial chromium(VI) concentration on metal adsorption. MWCNTs dosage: 10 g/L. Temperature: 20°C.

Equation		0.005 g/L	0.01 g/L	0.04 g/L
3	k (min^{-1})	0.48	0.27	0.26
	r^2	0.9008	0.7710	0.8968
4	k (min^{-1})	0.36	0.14	0.15
	r^2	0.9537	0.9739	0.9796
5	k (min^{-1})	0.11	0.06	0.06
	r^2	0.9004	0.8767	0.9760

Table 6.
The rate law governing the adsorption of chromium(VI) onto the MWCNTs.

experiments together with the experimental conditions used in the investigation were summarized in **Table 7**.

It can be concluded that, under the present experimental conditions, Dowex 1×8 resin is the most effective to remove hazardous chromium(VI) from near neutral or acidic solutions, whereas MWCNTs presented the worse registers. The above is not a bad conclusion about the use of these MWCNTs as adsorbents for Cr(VI), and not delegitimize the investigation presented in this work, only stated that there are other potential adsorbents-ion echangers that remove Cr(VI) from liquid effluents with a better efficiency.

3.7 Thermodynamics

Besides the data of the change of enthalpy (see Section 3.5) derived for the adsorption process of chromium(VI) onto the carbon nanotubes, a further thermodynamic analysis of the adsorption process can be considered taking into account the next equations:

$$\Delta G^{\circ} = -RT \ln b \tag{11}$$

$$\Delta G^{\circ} = \Delta H^{\circ} - T\Delta S^{\circ} \tag{12}$$

Accordingly with Eq. (11) in which b is the Langmuir constant showed in **Table 5**, it can be obtained that the value of ΔG° is 9 kJ/mol, confirming that the chromium(VI) uptake onto the nanotubes is nonspontaneous.

To calculate ΔS° for the present adsorption system, Eq. (12) is used, obtaining a value of −0.08 kJ/mol K. The negative value for the change of entropy characterizes a decrease disorder of the system when chromium(VI) is adsorbed onto the nanotubes.

Adsorbent-anion exchanger	Active group	pH 1	pH 4
Dowex 1x8	QAS-Cl^- form	89	>99
MWCNTs	None	32	42
ox-MWCNTs	–COOH	68	No data
Activated carbon	None	47	59

Aqueous solution: 0.01 g/L Cr(VI) at different pH values. Solid dosage: 1 g/L. Temperature: 20°C. Stirring speed: 1000 min^{-1}. Time: 1 h.
QAS: quaternary ammonium salt.

Table 7.
Percentage of Cr(VI) adsorption using various adsorbent-anion exchangers.

3.8 Chromium(VI) desorption

In the present investigation, the desorption of chromium(VI) from the metal-loaded carbon nanotubes was studied using hydrazine sulfate solutions as desorbent for the metal, at the same time, in the desorption process, Cr(VI) is reduced to the less hazardous Cr(III) oxidation state, accordingly to

$$4\,HCrO_4^- + 10\,H^+ + 3\,N_2H_4 \cdot H_2SO_4 \rightarrow 4\,Cr^{3+} + 16\,H_2O + 3\,N_2 + 3\,SO_4^{2-} \qquad (13)$$

Desorption experiments were carried out with aqueous solutions containing 25–50 g/L of hydrazine sulfate and 2 mg/g Cr(VI)-loaded MWCNTs at a 25 mL/g solution volume/weighed MWCNTs relationship and 20°C. The results from these experiments indicated that:

i. The variation in the hydrazine sulfate solution concentration has no effect on the reaction yield (95% chromium recovery from loaded nanotubes).

ii. The equilibrium is reached within 5 min of reaction.

iii. The desorbed solution contained a chromium(III) concentration near eight times the initial chromium(VI) concentration in the feed solution (0.01 g/L) of the adsorption experiments.

4. Conclusions

The adsorption of chromium(VI) onto the multi-walled carbon nanotubes is dependent on the pH values of the aqueous solution. The adsorption reaches a maximum at pH 4 and decreases at more acidic and alkaline pH values. The adsorption is exothermic ($\Delta H° = -14$ kJ/mol) and nonspontaneous (positive $\Delta G°$ valor), whereas at 20–60°C, the adsorption of chromium (VI) onto the nanotubes better fits to the pseudo-second-order model. In the 0.005–0.04 g/L range of chromium(VI) concentrations in the aqueous solution, the metal uptake onto the nanotubes responded well to the particle-diffusion model, and the metal adsorption responded to the Langmuir and Freundlich isotherms, indicating that the adsorption process is homogeneous and multilayer in nature. Chromium(VI) can be desorbed from MWCNTs by the use of hydrazine sulfate solutions, which releases to the aqueous solution chromium in the less hazardous (III) valence state.

Acknowledgements

This research was funded by the Ministry of Science, Innovation and Universities of the Spanish Government, in the “Challenges collaboration” call for proposals in 2017 (Ref. RTC-2017-6629-5).

Conflict of interest

The authors declare no conflicts of interest.

Author details

Francisco J. Alguacil and Félix A. Lopez*
National Center for Metallurgical Research (CENIM), Spanish National Council for Scientific Research (CSIC), Madrid, Spain

*Address all correspondence to: f.lopez@csic.es

References

[1] Alguacil FJ, Alonso M, Lopez FA, Lopez-Delgado A. Pseudo-emulsion membrane strip dispersión (PEMSD) pertraction of chromium (VI) using Cyphos IL101 ionic liquid as carrier. Environmental Science and Technology. 2010;**44**:7504-7508. DOI: 10.1021/es101302b

[2] Wang Y, Li Y, Zhong Y, Cui C. Chromium(VI) removal by tri-n-octylamine/n-heptane via a pseudo-emulsion-based hollow fiber strip dispersion technique. Desalination and Water Treatment. 2017;**63**:103-112. DOI: 10.5004/dwt.2017.20150

[3] Koubaissy B, Toufaily J, Cheikh S, Sayed Hassan M, Hamieh T. Valorization of agricultural waste into activated carbons and its adsorption characteristics for heavy metals. Central European Journal of Engineering. 2014;**4**:90-99. DOI: 10.2478/s13531-013-0148-z

[4] Li Y, Cui C. Extraction behavior of Cr(VI) and coexistence with interfering ions by trialkylamine. Journal of Water Reuse and Desalination. 2015;**5**:494-504. DOI: 10.2166/wrd.2015.110

[5] Bayramoglu G, Akbulut A, Yakup Arica M. Aminopyridine modified *Spirulina platensis* biomass for chromium(VI) adsorption in aqueous solution. Water Science and Technology. 2016;**74**:914-926. DOI: 10.2166/wst.2016.281

[6] Garza-González MT, Ramírez-Vázquez JE, García-Hernández MA, Cantú-Cárdenas ME, Liñan-Montes A. Reduction of chromium (VI) from aqueous solution by biomass of *Cladosporium cladosporioides*. Water Science and Technology. 2018;**76**:2494-2502. DOI: 10.2166/wst.2017.427

[7] Al Dwairi R. Modeling of chromium (VI) adsorption from aqueous solutions using Jordanian zeolitic tuff. Water Science and Technology. 2017;**75**:2064-2071. DOI: 10.2166/wst.2017.065

[8] Elyahyaoui A, Ellouzi K, Al Zabadi H, Razzouki B, Bouhlassa S, Azzaoui K, et al. Adsorption ofe chromium(VI) on calcium phosphate: Mechanisms and stability constants of surface complexes. Applied Sciences. 2017;7:222. DOI: 10.3390/app.7030222

[9] Wojcik G, Hubicki Z. Investigations of chromium(VI) ion sorption and reduction on strongly basic anion exchanger. Separation Science and Technology. 2018;**53**:1088-1096. DOI: 10.1080/01496395.2017.1335323

[10] Xing J, Shen Y, Yang B, Feng D, Wang W. A green method based on electro-assisted and photo-assisted regeneration for removal of chromium(VI) from aqueous solution. Water Science and Technology. 2018;**76**:896-902. DOI: 10.2166/wst.2018.260

[11] Naghizadeh A. Comparison between activated carbon and multiwall carbon nanotubes in the removal of cadmium(II) and chromium(VI) from water solutions. Journal of Water Supply: Research and Technology-AQUA. 2015;**64**:64-73. DOI: 10.2166/aqua.2014.022

[12] Alguacil FJ, Garcia-Diaz I, Lopez FA, Rodriguez O. Removal of Cr(VI) and Au(III) from aqueous streams by the use of carbon nanoadsorption technology. Desalination and Water Treatment. 2017;**63**:351-356. DOI: 10.5004/dwt.2017.0264

[13] Anastopoulos I, Anagnostopoulos VA, Bhatnagar A, Mitropoulos ACM, Kyzas GZ. A review of chromium removal by carbon nanotubes. Chemistry and Ecology. 2017;**33**:572-588. DOI: 10.1080/02757540.2017.1328503

[14] Xing J, Zhu C, Chowdhury I, Tian Y, Du D, Lin Y. Electrically switched ion exchange based on polypyrrole and carbon nanotube nanocomposite for the removal of chromium(VI) from aqueous solution. Industrial and Engineering Chemistry Research. 2018b;**57**:768-774. DOI: 10.1021/acs.iecr.7b03520

[15] Alguacil FJ, Lopez FA, Rodriguez O, Martinez-Ramirez S, Garcia-Diaz I. Sorption of indium (III) onto carbon nanotubes. Ecotoxicology and Environmental Safety. 2016;**130**:81-86. DOI: 10.1016/j.ecoenv.2016.04.008

[16] Alguacil FJ, Escudero E. The removal of toxic metals by ion exchange resins. Part VIII: Arsenic(III)/OH^-/Dowex 1x8. Revista de Metalurgia. 2018;**54**:e132. DOI: 10e.3989/revmetalm.132

[17] Alguacil FJ, Alcaraz L, Garcia-Diaz I, Lopez FA. Removal of Pb^{2+} in wastewater via adsorption onto an activated carbon produced from winemaking waste. Metals. 2018;**8**:697. DOI: 10.3390/met8090697

[18] Largitte L, Pasquier R. A review of the kinetics adsorption models and their application to the adsorption of lead on an activated carbon. Chemical Engineering Research and Design. 2016;**109**:495-504. DOI: 10.1016/j.cherd.2016.02.006

[19] Diaz-Pavon AL, Cerpa A, Alguacil FJ. Processing of indium(III) solutions via ion exchange with Lewatit K-2621 resin. Revista de Metalurgia. 2014;**50**:e010. DOI: 10.3989/revmetalm.010

[20] Alguacil FJ. The removal of toxic metals by ion exchange resins. Part III: Cu^{2+}/sulphate/Amberlite 200. Revista de Metalurgia. 2003;**39**:205-209. DOI: 10.3989/revmetalm.2003.v39.i3.330

Chapter 5

Advances in Metal Recovery from Wastewaters Using Selected Biosorbent Materials and Constructed Wetland Systems

John G. Murnane, Bashir Ghanim, Lisa O'Donoghue, Ronan Courtney, Thomas F. O'Dwyer and J. Tony Pembroke

Abstract

An expanding global population not only increases the amounts of municipal solid waste and wastewater generated but also raises demand for a wide range of raw materials used to manufacture goods. Extraction of these raw materials and many subsequent manufacturing processes contribute significantly to the presence of a variety of metals in wastewaters and leachates. Metal-rich wastewaters not only result in short- and long-term environmental and associated health concerns but also have potential economic value if the metals can be recovered. In this chapter, we review the effectiveness of biochar, microbial and lignin biosorbents as well as constructed wetland systems to remove soluble metals from wastewaters. The wide variation in adsorptive capacity of these biosorbent materials reflects the heterogeneous nature of the source materials used for their production. Physical and chemical modifications of biochars and lignins generally improve their adsorptive capacities which remain highly variable. Constructed wetlands are attractive because of their passive nature with low-energy and low-maintenance requirements, although their long-term capacity to treat metal-rich wastewaters is as yet largely undetermined. Future perspectives focus on increasing the selectivity of adsorbents to remove complex matrices of metals from wastewaters and on increasing their adsorption/desorption capacities.

Keywords: wastewater, biosorbent materials, biosorption, biochar, lignin, microbial adsorption, constructed wetlands

1. Introduction

Metals such as lead (Pb), nickel (Ni), silver (Ag), aluminium (Al), cadmium (Cd), zinc (Zn), chromium (Cr), copper (Cu), vanadium (V), platinum (Pt), mercury (Hg) and titanium (Ti) are found in wastewaters associated with many industrial processes. Such processes include milling, etching, electroplating, conversion-coating, electrolysis and waste-to-energy facilities to mention but a few. Certain pigment manufacturing processes utilise Cr and Cd, while Cu and arsenate are utilised during wood processing. The petroleum industry utilises significant amounts of catalytic material which includes V and Ni, while mining operations produce a variety of extractive residues and leachates covering a very wide range of metals [1, 2]. Such industrial processes contribute significantly to the presence of metals in wastewaters. In addition, municipal solid waste contains significant quantities of metals such as silicon (Si), calcium (Ca), iron (Fe), Al, sodium (Na), magnesium (Mg), Zn, Cu and Pb [3] and can result in a highly toxic landfill leachate whether deposited as untreated waste or as incinerated ash residue.

Public interest in metal contamination of wastewater has some of its origins in the use of Pb as a gasoline additive in the United States which sparked a general interest in heavy metals as potential hazards in the minds of the public [4]. In addition the general low levels at which metals demonstrate toxicity is another key issue. In nature, heavy metals occur mainly as insoluble forms in natural mineral deposits occurring as silicates, carbonates, oxides or sulphides, which in general weather slowly. However not all are slow weathering, and as an example rainwater may solubilise rocks containing magnesium oxides as magnesium bicarbonate due to the carbon dioxide content of rainwater, whereas Fe may also dissolve but precipitates as insoluble ferric hydrate [4]. Many heavy metals precipitate in a similar way to Fe and thus are rarely present at neutral pH. With an increased awareness of metals as a component of wastewater and leachates and the often low level at which toxicity is demonstrated, there has been a keen interest in both their environmental and health-related consequences.

As all humans on the planet need drinking water, there is a vested interest in ensuring that it is of the best quality possible. The source of most raw water used for potable consumption is abstracted from surface waters (i.e. lakes and rivers), and these in turn are impacted by industrial, municipal as well as agricultural wastewater discharges. Most developed countries have drinking water quality standards. In Europe these are guided by the European Drinking Water Directive [5], in the United States the US Environmental Protection Agency (USEPA) utilise the Safe Drinking Water Act [6] and in Australia the Australian Drinking Water Guidelines are used [7]. Other countries use World Health Organisation (WHO) guidelines [8]. All of these guidelines specify maximum metal concentrations considered safe for human consumption, a selection of which are included in **Table 1**. It is critical therefore that

Parameter	Drinking water regulations (μg L^{-1})			
	EU	United States	Australia	WHO
Al	200	N.S.	200	N.S.
As	10	10	7	10
Ba	N.S.	2000	700	1300
Cd	5	5	2	3
Cr	50	100	50	50
Cu	2000	1300	2000	2000
Hg	1	2	1	6
Ni	20	N.S.	20	70
Pb	10	15	10	10

N.S.—not specified.

Table 1.
Selection of specified maximum metal concentrations (μg L^{-1}) for drinking water in the EU [5], the United States [6], Australia [7] and the WHO [8].

Metal remediation technique	Basis of process
Precipitation, coagulation and flocculation	Precipitation was one of the earliest methodologies to treat metal wastewater. Lime precipitation (or indeed on occasion limestone) is often a first treatment used to remove metals particularly from acidic metal-containing wastewaters. Precipitation with lime, containing predominately calcium oxides and hydroxides, is one of the cheapest and simplest techniques with precipitation of the metal species such as Zn, Cu, Fe, Mn, Co and Ni as hydroxides. Incomplete precipitation can occur for other metals such as Cd, Pb and Hg, and in such cases soda ash can be employed based on its carbonate ion to precipitate Pb, while sodium sulphide has been used for Cd and Hg [19]. However these may give rise to the production of relatively large quantities of toxic sludge and incomplete removal [9, 20, 21]. Other agents such as alum, ferric chloride and a variety of polymers can also be used to flocculate and precipitate metals [22, 23]
Electrodeposition	Many industrial processes contain acid solutions saturated with metals, such as Cu, which may be ideal for the use of electrochemical techniques such as electrodeposition of the metal onto the cathode surface. Since the electron is the main reagent of the reduction reaction, the electrochemical process can be considered a clean technology and environmentally attractive [24]
Ion exchange	Traditionally developed to recover uranium (U), ion-exchange methods have widespread applicability for metal recovery with the use of a variety of resins to remove solubilised heavy metals from a variety of sludges [25]
Solvent extraction	Here specific organic solvents can be utilised to extract metal ions as an organic solvent soluble form. This is then recovered by acid treating the organic solution causing the metal to be recovered in a concentrated form [26]
Cementation	Cementation is a precipitation process whereby ions are reduced to zero charge at a solid metallic interface. A key example is where Cu ions in solution from ore leachate are precipitated in the presence of Fe. As the Fe oxidises, the Cu reduces with the Cu recovered on the surface of the Fe. This process can be used for several metals such as Zn or Cd [12]
Reverse osmosis	This technology pressurises water to enable it to pass from an area of high to low solute density through a semipermeable membrane which has a defined pore structure and size. This is the reverse direction to which natural osmosis occurs and is thus termed reverse osmosis. The semipermeable membrane captures the solutes as the water passes through. This technology is applied widely for desalination but can also have a role to play in metal recovery from wastewaters [27]
Ultrafiltration	Ultrafiltration is a membrane-based technology which utilises pressure to separate material through semipermeable membranes. High-molecular-weight materials are retained, while water and low-molecular-weight materials permeate the membranes. The retentate characteristics are a function of the molecular weight and exclusion size of the membranes. The technique combines complexation and ultrafiltration where soluble metal-binding polymers are added to complex the metals which then become concentrated and bind to the polymeric membrane material [17, 18]. Metal-binding ligands include carboxymethyl or diethylaminoethyl celluloses, chitosan, polyvinyl alcohols and polyacrylic acid [18, 28]
Adsorption processes	A range of adsorbent materials have been used to remove metal species from waste waters. Such materials include mineral organic materials, activated carbon, zeolites, wastes, biomass or natural or synthetic polymeric materials [29, 30]. Certain applications can be termed sorption flotation depending on the technology roll out [11, 31]. A small number of applications have emerged where microorganisms rather than polymeric materials have been utilised with adsorption to natural polymeric materials associated with the surface of microorganisms [2, 20]

Table 2.
Common technologies utilised to remove metals from wastewater streams.

wastewater treatment technologies have the capacity to remove or reduce effluent metal concentrations prior to discharge to receiving waters, such as rivers and lakes.

Within the range of treatment methods, there are a variety of technologies based on physiochemical methodologies. These include chemical precipitation [9], coagulation-flocculation [10], flotation [11], cementation [12] and electrocoagulation [13]. Adsorption onto zeolites, clay and resin ion exchange [14, 15] and membrane filtration techniques such as ultrafiltration, nanofiltration and reverse osmosis [16–18] have also been used (**Table 2**). Adsorption approaches have largely focused on the use of zeolites.

There is increased interest in the removal of metals from wastewater, not only because of expansion in industrial sectors which produce metal waste streams but also from mining activity, which is subject to rigorous regulation. In addition there is the added incentive to recover metals from such streams that may have unexpected economic value. Clearly a range of current technologies exist for metal removal from waste streams; however many of these require significant costs and the use of high-end technologies. Other cheaper adsorption options have been based, in large measure, around the use of activated charcoal and ion-exchange resins; however, in recent years there has been a developing focus on the potential use of biosorbent materials for metal removal from waste streams. These biosorbents possess a number of useful advantages in that they are, in most cases, naturally occurring, cheap, readily available in large quantities and can generally be modified to act as effective adsorbents for a range of metals from wastewater. The following sections of this chapter focus specifically on recent research in the application of some selected biosorbents including biochars, lignins and microbes and on constructed wetland systems in the removal of metals from wastewaters.

2. Bioremediation of metals from leachates

2.1 Biochars

2.1.1 Nature, sources and production of biochar

Biochar is a low-cost carbonaceous material derived from the thermal conversion of various biomasses using techniques such as gasification [32], pyrolysis [33–36], hydrothermal carbonisation [37] and torrefaction [38], at temperatures ranging from 300 to 900°C and in oxygen-limiting environments. High pyrolysis temperatures in the carbonisation of biomass (>500°C) lead to high surface areas, microporosity and a biochar that is highly hydrophobic in nature [33, 37]. Low pyrolysis temperatures (<500 °C) lead to partial carbonization and the presence of more oxygen-containing functional groups, lower surface areas and a biochar with more affinity for binding inorganic species from solution [33, 37]. Depending on the different types of thermal conversion approach, biochars can be designed to display a range of properties which make these materials suitable for pollutant removal scenarios. In particular, these properties include variable surface area, microporosity, surface charge and pH, polarity, adsorption and ion-exchange capacity [36]. A selection of typical biochar feedstocks and the associated compositions of their respective biochars, after formation, are outlined in **Table 3**. Typically biochars have been produced from woody-type wastes [32, 33], manures [37, 38], agricultural wastes [36] and energy crops such as alfalfa [34] and miscanthus [39]. Biochars derived from wood or crop wastes typically tend to show higher surface areas, whereas animal waste and activated sludge-derived biochars tend to exhibit lower surface areas.

	Proximate analysis (%)			Ultimate analysis (%)				
Feedstock	**Volatile matter (%)**	**Fixed carbon (%)**	**Ash content (%)**	**C**	**H**	**N**	**O**	**Reference**
Oak sawdust	69.2	16.5	0.81	52.3	5.7	0.06	41.9	[32]
Pine sawdust	83.1	16.8	0.10	51.0	6.0	0.10	42.9	[33]
Alfalfa	78.9	15.8	5.3	49.9	6.3	2.80	40.8	[34]
Bamboo	81.6	17.5	0.9	52.0	5.1	0.40	42.5	[35]
Corn straw			60.2	35.9	1.6	0.43	1.9	[36]
Poultry litter			37.7	49.8	4.4		3.2	[37]
Pig manure	19.1		46.5	44.1	2.5	2.1		[38]
Giant miscanthus	65.3	15.6	11.7	46.2	6.0	—	45.9	[39]

Table 3.
Typical biochar feedstocks and associated composition.

2.1.2 Modification and activation of biochars

Following production of biochars, a physical or chemical activation process can be used to enhance the material surface area and pore fraction or simply to form surface functional groups, all of which can enhance the material's ability to function as an adsorbent for metal uptake. Specific physical activation methods are mostly based around the use of steam [40, 41], while chemical activation can be accomplished with the use of either base/oxidant [42] or acid/oxidant [43] combinations post pyrolysis or with the incorporation of metals pre-pyrolysis [44]. Typical activation methods are outlined in **Table 4**, and almost all of these methods have resulted in enhanced metal uptake from solution.

2.1.3 Application of biochars for metal recovery

Thus far much of the work on the use of modified and unmodified biochars has been aimed at the recovery of many of the transition and heavy metals such as arsenic (As) [45], Cr [46, 47, 52], Cu [37, 41, 51], Pb [49, 53, 56], Cd [40, 48], Zn [36], Ni [50], Hg [54] and U [57] from selected waste streams. Examples of the relative uptake of these metals by selected biochars can be seen in **Table 5**. The influence of solution pH on metal uptake levels varies significantly with many of the metals exhibiting maximal uptake in the range pH 4–8. The surface charge on the adsorbent and the solution pH are important. At low pHs, it is likely that the biochar surface is protonated and may present a suitable binding opportunity for metals in their anionic or negatively charged form. Increasing the solution pH can lead to a reduction in protonation of the biochar surface and a greater opportunity for the metal in its more cationic state to bind to the biochar.

Many of the metals, at strongly acidic conditions, will exist in their cationic states, but once the solution pH rises to between pH 5 and pH 8, many of these metals can be precipitated as hydroxide species and as such become unavailable for adsorption. Hence acid pHs tend to favour adsorption onto biochar materials. At pH 2 and less, there appears to exist significant competition for adsorption sites between the M^{n+} form of the cation and H^+ in solution. As the pH rises to between pH 2 and 5, H^+ concentration decreases, leading to less competition with the M^{n+} form of the cation for adsorption sites on the biochar. Metal uptake tends to move significantly towards its maximum uptake level in this latter pH range. Beyond pH 5 the cationic form of the metal starts to shift towards a hydroxylated species in solution. Once moving towards pH 7, it can be difficult to determine whether adsorption or simply precipitation is taking place onto the biochar. The variations in metal form and biochar surface charge arising from variation in solution pH may potentially be of significant benefit in the regeneration of biochars.

Modification of some of these biochars by the broad methods outlined in **Tables 4** and 5 can, in many instances, lead to an enhancement of metal uptake levels. For example, adsorption of hexavalent chromium from aqueous solution was shown to rise significantly, with the presence of an increased number of amino groups being suggested to significantly enhance metal uptake onto a polyethyleneimine modified rice biochar [52]. Other scientific explanations for increased uptake of selected metals following modification of specific biochars, as outlined in **Tables 4** and 5, include how a modification of peanut shell led to an increased specific surface area [53]; how treatment of corn straw biochar with sodium sulphide yielded more oxygen-containing functional groups on the surface [54]; how larger pore sizes, pore volumes, and more functional groups could be achieved with the treatment of wheat straw biochar with graphene oxide [55]; and how higher contents of surface carboxylate groups and ultimately negative surface charge on the modified biochar could be achieved with nitric acid treatment of cow manure biochar [57].

2.2 Microbial

There have been many studies on the effects of metals on soils and water in the natural environment and particularly on the microflora that interact with these metals [59, 60]. Such studies have revealed that soil and natural water streams contain a range of microorganisms with the capability of metal transformation. Such effects can be exploited not only in soil remediation but also in utilising such microorganisms as tools for remediation of wastewaters contaminated with metals. There are currently several categories of interaction that can be observed between microbial populations and metal species as summarised in **Table 6**.

Feedstock	Modifying agent	Nature of modification	Stage			Reference
Poultry manure	Steam	Physical	Post-pyrolysis	Cd^{2+}	+	[40]
Pine sawdust	Steam	Physical	Post-pyrolysis	Cu^{2+}	=	[41]
Municipal sludge	Bases	Chemical	Post-pyrolysis	As(V)	+	[42]
Peanut hull/ hydrochar	Acid/ oxidant	Chemical	Post-hydrothermal treatment	Pb^{2+}	+	[43]
Pine wood/ $MnCl_2$	Metals	Chemical	Pre-hydrolysis			[44]

Table 4.
Biochar activation methods.

Metal	Feedstock/modifier	Optimum pH range	Sorption capacity (mg g^{-1})	Reference
Unmodified biochars				
As(III)	Rice husk	8.0	19.3	[45]
Cr(III)	Peanut stalk	4.0	25.0	[46]
Cr(VI)	Sugar beet tailing	2.0	123.0	[47]
Cd(II)	Dairy manure	—	31.9	[48]
Pb(II)	Sludge	5.0	30.9	[49]
Cu(II)	Hardwood	5.0	6.8	[36]
Zn(II)	Corn straw	5.0	11.0	[36]
Ni(II)	Almond shell	6.0	20.0	[50]
Cu(II)	Pig manure	5–6	75.49	[51]
Modified biochars				
Cd(II)	Poultry manure/steam	—		[40]
As(III)	Rice husk/Fe	—	30.7	[45]
Cr(VI)	Rice husk/ polyethyleneimine	2.0–7.0	436.0	[52]
Ni(II)	Peanut/KOH	6.0–7.0	87.1	[53]
Hg(II)	Corn straw/Na_2S	4.0–6.0	5.7	[54]
Hg(II)	Wheat straw/graphene oxide	6.8–7.0	0.85	[55]
As(V)	Pine wood/Mn	8.0	6.5	[56]
U(VI)	Cow manure/HNO_3	4.5	355.6	[57]
Cu(II)	Pine sawdust/H_3PO_4	—	30.0	[58]

Table 5.
Typical metal adsorption levels for unmodified and modified biochars.

Fungal and bacterial species that can grow and metabolise in the presence of metal species have been demonstrated to possess a variety of mechanisms of resistance that in many cases have a genetic basis [67–69]. The nature of the microbial species and the mechanism used are often dependent on the metal species present, and several mechanisms have been identified [67]. These include exclusion of the metal species by production of an extracellular matrix that acts as a barrier to entry of the metal. Such material known as extracellular polysaccharide substance (EPS) binds metal species and can have biotechnological applications [2]. Other techniques involve active metal efflux which is often associated with microbial mobile genetic elements [70], sequestration of metals, enzymatic detoxification exemplified by Hg resistance mechanisms and reduction of target sensitivity of the microorganism [67]. Understanding such mechanisms offers insights into the methodologies that have evolved in biological systems over millions of years and may offer new biotechnical approaches that can be exploited for metal remediation.

2.2.1 Microbial adsorption mechanisms

Microbial EPS material is often associated with the formation of stress-responsive structures on the surface of microbial cells during biofilm formation, and thus its production can be adaptive or protective in nature [2]. EPS can exist in many forms and can be genera specific in its chemical composition. In general

Type of microbe-metal interaction	Reference
Enzymatic transformation of the metal species such as the use of thiosulphate reductase to transform thiosulphate to sulphite for precipitation of metals such as Cd or Zn as metal sulphide species	[61, 62]
Accumulation of metal intercellularly. Examples include Ni accumulation as its phosphide and carbide crystal in *Pseudomonas aeruginosa*	[63]
Cell appendage adhesion such as metal binding to cell fimbriae as with the *E. coli* FimH adhesion binding to Pb, cobalt (Co) and Cr	[64]
Bioadsorption of metal species to microbial surface polymers. Many microbes produce surface polymeric materials as stress responses in the environment to resist change as protection against environmental stresses. Many of these extracellular polysaccharide substances are highly charged materials that bind metal species	[2, 29]
Bioleaching activities where metal species may be solubilised as a result of acid production by the interacting microbial species. Examples include leaching by sulphur-oxidising bacteria through production of sulphuric acid followed by precipitation of insoluble sulphides by sulphate-reducing microorganisms in acid environments. Citric acid produced by *Aspergillus niger* has also been utilised in alkaline environments to leach metals from alkaline wastes and soil	[65, 66]

Table 6.
Interaction between microbial populations and metal species.

EPS consists of repeating monosaccharide units forming hetero- or homopolymers linked in glycosidic linkage which in turn are linked to a lipid anchor. The complex composition and branching of the EPS with many phosphate, amino sugars and hexuronic acid residues [2, 71] give the EPS structure an anionic charge which can be utilised to protect the microbe from metal toxicity. Key chemical groups on microbial surfaces include carboxyl groups associated with microbial peptidoglycan, phosphate groups on surface material, uronic acid and charged amino acid groups [2].

There have been many studies of the utility of microbial cells in metal binding [72] although few actual processes have yet been developed. Cyanobacteria and algae have been extensively studied to bind a variety of toxic metals in natural water systems in polluted environments [73–76]. The rationale for their use is that many cyanobacterial and algal species are normal inhabitants of these water courses, and hence their presence would be less environmentally intrusive. Biosorption of a variety of metals has been studied including antimony (Sb), Ni, Cd, Cu, Pb, Co, Mn, As and Zn [2]. A rotating biological reactor containing microbial biofilms has previously demonstrated that metals can be accumulated in a process environment [77], and while this is as yet one of the few processes that have been developed, the potential to utilise microbial biomass is evident. Immobilised photosynthetic bacteria have been utilised to remediate swimming pool muds in Fukushima following the radioactive leak as a result of a Japanese tsunami [78] demonstrating the potential practicality of utilising microbes for metal remediation, in this case radioactive nuclides.

Table 7 outlines a number of microbes which have been tested for their metal biosorbent capabilities. An uptake of 277.5 mg Pb(II) g^{-1} from aqueous solution at an optimum pH of 6.8 was measured in a study using the exopolysaccharides from *Paenibacillus peoriae* strainTS7 [79]. Another study which used the biosorbent *Anabaena doliolum Ind1* has outlined the presence of a range of surface groups including carboxyl, carbonyl, hydroxyl, amides and sulphate groups as being key binding sites for metals such as Cd(II) [80]. In a study which examined equilibrium and kinetic and thermodynamics of aqueous Al biosorption by *Streptomyces rimosus* biomass, the presence of methyl, hydroxyl, amine, carboxyl, thiol and phosphate groups was identified as significant binding sites [81], while fatty acid, amide, lipids

Strains	Source	Heavy metal	Optimum pH	Uptake level ($mg\ g^{-1}$)	Reference
Paenibacillus peoriae strainTS7	Aqueous solution	Pb(II)	6.8	277.5	[79]
Anabaena doliolum Ind1	Contaminated soil coal mine	Cd(II)	7.0	92% (2 ppm)	[80]
Streptomyces rimosus	Antibiotic manufacturer	Al(III)	4.0	11.7	[81]
Providencia vermicola strain SJ2A	Soil (battery manufacturing site)	Pb(II)	—	155.1	[82]
Rhizobium radiobacter strain VBCK1062	Contaminated soil fertiliser/chemical industry	As(V)	—	0.068	[83]
Rhodococcus opacus	Water streams	Al(III)	5.0	41.6	[84]

Table 7.
Microbes used for extracellular polysaccharide substance (EPS)-assisted heavy metal removal from selected wastewaters.

and protein moieties were identified as significant contributors to the adsorption of Pb(II) on *Providencia vermicola strain SJ2A* [82]. A study to measure the biosorption of Al(III) from waste streams using a *Rhodococcus opacus* strain reported a removal rate of 41.6 mg g^{-1} and indicated that the presence of amine, alkyl, carbonyl and phosphate surface functionalities were significant in its removal [84]. There are several potential advantages of utilising microbial biosorption. They can be classed as environmentally safe and do not generate toxic by-products, while in addition both live and dead cells can be used. On the downside however, mild desorption must be used to recover metals, while the efficiency of biosorption may be low and dependent on sensitive physiochemical parameters. In addition metals may affect the viability of adsorbing strains and limit reuse and cycling [85].

2.3 Lignins

2.3.1 Sources and production of lignin

Lignin is a natural biopolymer which makes up typically 15–30% of wood and grass biomass and provides structural rigidity to many plant cell walls. Its structure depends largely on the plant type, age and growth location [86]. Wood-based lignins in particular have been used for adsorption of air pollutants, organics and heavy metals due to their physicochemical properties, low cost, abundant availability and extent of active adsorption sites [87]. Separation and isolation of lignin from cellulose is generally quite difficult due to condensation and oxidation reactions that occur during the separation process which generally consist of either chemical (e.g. alkaline pulping or acid hydrolysis) or mechanical separation processes (e.g. ball milling). In industry, significant quantities of lignin (approximately 70 million tonnes in 2017) are produced as a by-product of the 'kraft' paper manufacturing process. In this process woodchips are treated with sodium hydroxide (NaOH) and sodium sulphide (Na_2S) to separate lignin from the cellulose fibres of wood which results in a black-coloured lignin-rich liquor. Several efforts to extract lignin from the liquor and use it as an adsorbent of inorganic pollutants, including metals, have been made with varying degrees of success [88, 89].

2.3.2 Unmodified lignin as an adsorbent

Lignin polymers are hydrophobic in nature containing carboxyl, hydroxyl and phenolic surface groups, which give them an affinity for metal ion adsorption. Their specific surface area is relatively low, typically around 100–200 $m^2\ g^{-1}$ [90]. Similar to biochars, the absorption research focus to date has been on heavy metals such as Cu, Cd, Cr, Pb and Zn, with typical adsorption rates using unmodified lignin-based adsorbents between 4.2 mg Cu g^{-1} using straw-based lignin and 137 mg Cd g^{-1} using lignin from kraft liquor (**Table 8**). The pH of the solution significantly influences metal uptake as described in the previous section on biochars, with generally higher adsorption rates found at pH 5–6.

Unmodified lignins tend to have relatively low aqueous metal adsorption capacity and in addition have poor selectivity for certain metals [94]. Because of this, modification and activation is frequently carried out on lignins to improve their metal adsorption properties.

2.3.3 Modification of lignins to enhance adsorption

Lignins can be chemically modified to improve their physicochemical properties, in particular their adsorption capacity, hydrophobicity and hydrophilicity as well as their overall stability. The modifications are usually carried out with oxygen-, nitrogen- or sulphur-containing functional groups which react mostly with the primary phenolic hydroxyl groups. For metal adsorption, the principal functional groups are divided into three categories [94]. The first category is oxygen-containing functional groups where acid treatment increases the naturally occurring fraction of oxygen groups as well as the hydrophobicity of the lignins. This process can significantly increase the adsorption rates for metals such as Cd and Pb ([95], **Table 9**) provided the pH is also controlled. In general, oxidised lignins have been shown to exhibit stronger adsorption capabilities than unmodified lignins due to the higher amounts of carboxyl groups present [96]. The second category is nitrogen-containing functional groups such as amines and triazoles, which have a high affinity for soluble metals and can be grafted onto the lignin by a Mannich reaction [97]. It has been shown that nitrogen-modified lignin increased the adsorption capacity for Pb(II) by over four times that of the original lignin [98].

Lignin type	Maximum metal adsorption capacity (mg g^{-1})						Reference
	Cu(II)	Cd(II)	Cr(III)	Pb(II)	Zn(II)	pH	
Wheat straw	4.2					6.0	[90]
Wheat straw	26.0					6.0	[91]
Kraft liquor (eucalyptus pulping)	87.1	137.1				4.5	[89]
Beech wood		6.7		8.2		5.0	[92]
Poplar wood		7.5		9.0		5.0	[92]
Kraft liquor from paper mill			18.0			5.0	[88]
Kraft liquor from paper mill					73	5.0	[93]

Table 8.
Typical metal adsorption capacities of unmodified lignins.

Functional group type	Lignin source	Maximum metal adsorption capacity (mg g^{-1})					Reference
		Cu(II)	Cd(II)	Cr(III)	Pb(II)	pH	
Oxygen-containing	Wheat straw		35.9		155.4	5.0	[95]
	Wheat straw	399.0				5.5	[99]
	Sugar cane bagasse				107.5	6.0	[100]
	Sugar cane bagasse		67.7			5.0	[100]
	Lignosulfonate	59.9	48.8		194.5	6.0	[101]
	Lignosulfonate			41.8		5.0	[101]
	Cedar wood powder		129.3		370.8	5.2	[102]
Nitrogen-containing	Industrial black liquor	72.5			55.4	6.5	[103]
	Alkaline black liquor				60.5	6.0	[99]
Sulphur-containing	Industrial black liquor				64.9	5.0	[104]
	Alkaline black liquor	175.9			103.4	6.0	[105]
	Bamboo		72.4			6.0	[106]
	Industrial black liquor				188.0	5.0	[107]

Table 9.
Typical metal adsorption capacities of modified lignin.

The third category is sulphur-containing functional groups which have a strong affinity to metal ions such as Cd, Cu, Pb, Hg and others and are therefore used to enhance the adsorption properties of lignin (**Table 9**).

2.3.4 Activation of lignins to enhance adsorption

Lignin is one of a number of source materials used to manufacture activated carbons, which are commonly used in the water industry for removal of dissolved organic and inorganic pollutants. Activated carbons normally come it two types, granular activated carbon (GAC) and powdered activated carbon (PAC). GAC is normally used as a tertiary filter in water treatment processes and typically comprises particles of size 0.5–1.5 mm. The operation of a GAC filter is similar to that of a sand filter where GAC can be regenerated once it reaches a particular saturation ratio. PAC on the other hand comprises smaller particles, typically <0.2 mm, which are added to the water as an adsorbent. PAC tends to have very high adsorption rates due to its high specific surface area, but unlike GAC, it is very difficult to regenerate because of the difficulty in recovering the PAC powder from the water. Lignin-derived activated carbons can have very high adsorption rates because of their high micropore volume of up to 1 cm^3 g^{-1} and large specific surface area in the range 500–2000 m^2 g^{-1} [108, 109]. Preparation of lignin-based activated carbon can be a two-step physical process comprising carbonation and activation or a one-step chemical process. In the two-step physical process, carbonation is achieved by pyrolising the lignin in an inert atmosphere at temperatures in the range 600–900°C, which results in the formation of a char. Activation then follows by further heating to higher temperatures of 700–1200°C during which the porosity is developed using agents such as CO_2, N_2, air or steam. This activation step increases the surface area and pore volume by removing internal carbon mass and volatile organic residues. Chemical activation on the other hand normally uses chemical agents such as NaOH and KOH to impregnate the lignin which is then pyrolised at temperatures of 500–900°C in an inert atmosphere. In this process carbonisation and activation occur simultaneously resulting in activated carbon with an open porous structure and high specific surface area [110].

Typical adsorption rates of heavy metals from water, using wood-based activated carbons, range from 5.7 mg Cu g^{-1} using rubberwood sawdust to 255 mg Zn g^{-1} using oakwood (**Table 10**) and are influenced by the pH of the solution. For example, in a study to measure Cu(II) adsorption onto activated carbon-derived from rubberwood sawdust, optimum adsorption was measured at pH 6; however, at pH < 5, uptake decreased because of competition between H^+ protons and free Cu(II) ions to the fixation sites [111]. Similarly, optimum adsorption of Cu, Ni and Zn onto a wood-based activated carbon powder was observed at pH 6, decreasing at pH < 6, while metal precipitation was observed at pH > 6 [112].

Unlike unmodified lignins, modified and activated lignins demonstrate a high adsorption capacity for metals and are considered a promising biotechnology for their adsorption from wastewater. Although abundantly available, one of the difficulties with lignin is its heterogeneity and therefore its ability to consistently produce suitable adsorbents. Lignin quality is very much dependent on its source as well as the processes used to isolate it, and these and other factors very much influence its metal adsorption capacity and selectivity.

2.4 Constructed wetlands

2.4.1 Constructed wetlands for wastewater treatment

The term constructed wetland (CW) refers to a technology designed to employ ecological processes found in natural wetland ecosystems. Constructed wetland systems utilise wetland plants, soils and associated microorganisms to remove contaminants from wastewater and are gaining popularity due to low operating

Activated carbon feedstock	Metal adsorption capacities of wood-based activated carbons (mg g^{-1})						Reference
	Cu(II)	Cd(II)	Cr(VI)	Pb(II)	Zn(II)	pH	
Rubber wood sawdust	5.7					6.0	[111]
Moringa oleifera wood	11.5				17.7	6.0	[112]
Wood apple shell		27.6				6.5	[113]
Rubber wood			44.1			2.0	[114]
Fir wood			180.3			3.0	[115]
Acacia mangium wood			37.2			2.0	[116]
Tamarind wood				134.2		6.5	[117]
Tamarind wood				43.9		6.0	[118]

Table 10.
Typical metal adsorption capacities of wood-based activated carbons.

costs, reduced energy requirements, low maintenance and enhanced environmental benefits [119]. Constructed wetlands are passive treatment systems and may be broadly categorised in terms of (i) hydrology (surface/subsurface flow), (ii) flow path (horizontal or vertical flow) and (iii) type of macrophytic growth (free floating, submerged or emergent plant growth) [120]. The two most commonly used types of CW are the free water surface (FWS) wetland and vegetated submerged bed (VSB) wetland, also termed horizontal subsurface flow wetlands. The FWS CW is a shallow wetland with a combination of emergent aquatic plants (bulrush, reeds and others), floating plants (duckweed, water hyacinth and others) and submergent aquatic plants (pondweed, widgeon grass and others). An FWS CW may have open-water areas dominated by submergent and floating plants and may contain raised habitat areas. The main treatment processes of FWS CWs include sedimentation, as well as biochemical and physical transformations. The VSB operates differently from the FWS wetland in that emergent plants are rooted in gravel through which the wastewater flows. This system is also shallow, and the gravel size is sufficiently large to facilitate long-term subsurface flow without clogging. Roots and tubers (rhizomes) of the plants grow into the pore spaces of the gravel which are mostly anoxic due to permanent saturation, although local aerobic zones exist around the plant rhizomes which create an area of complex biochemical activity.

Wetland characteristics such as the soil medium, vegetation community and microbial populations influence processes such as deposition and filtration [121], while other processes such as oxidation, reduction, adsorption and precipitation further remove contaminants [122]. Wetland vegetation has a series of roles in wetland systems including transpiration, water baffling, sediment retention, provision of habitats to microorganisms and enhancing the residence time of the wetland system [123, 124].

2.4.2 Use of constructed wetlands for metal removal

The application of constructed wetlands as passive remediation systems for low pH effluent is well documented [119]. Indeed application of wetland technology has demonstrated potential for treatment of a wide range of pollutants associated with mining and mine processing where remediation is undertaken using only naturally available energy sources such as microbial metabolic energy, photosynthesis and topographical gradient [125]. CWs are capable of biosorption, metal sulphide redox transformations and microspecies plant interactions when treating metal-rich waters [126, 127]. More recently the potential for wetland ecosystems and CWs to buffer high pH effluents has received attention [119, 128].

CWs can be effective in high Al removal rates (>90%) due to formation of insoluble compounds through hydrolysis and/or oxidation which leads to the formation of a variety of oxides [128–131]. For example, immobilisation and attenuation of V by sorption to readily extractable oxides and carbonate phases in soils were previously measured [128]. High Al removal was also reported in VSB wetlands [129–131], and this was also attributed to the formation of insoluble compounds through hydrolysis and/or oxidation leading to the formation of a variety of oxides.

A selection of typical heavy metal uptake rates from wastewater using CWs and their primary modes of removal are shown in **Table 11**.

In a study to measure removal of heavy metals from industrial wastewater, it was reported that with the exception of Ni, the sediment concentrations for all metals decreased as the distance between the inlet and outlet increased. In addition some of the macrophytes had a higher uptake of metals than others, in particular *Typha latifolia* (cattail), which demonstrated a high uptake of all metals, in particular Ni. The average proportion of metal uptake by the macrophytes was 79% (roots 56%, aerial 24%) compared with 21% for the sediment [132].

Removal medium	Metal concentrations in sediments, roots and aerial tissues (mg kg^{-1})						Reference
	Cu	Cd	Cr	Pb	Fe	Ni	
Sediment	1.5–2.7	0.7–1.8	0.4–1.3	1.0–2.9	1.0–1.7	2.2–2.9	
Root tissue	3.9–5.6	2.3–5.2	1.4–2.3	3.8–7.2	2.7–5.7	4.3–7.6	
Aerial tissue	1.3–2.7	0.9–2.5	0.4–1.2	1.5–3.2	1.0–2.8	1.8–4.0	[132]
Total	6.7–11.0	3.9–9.5	2.2–4.8	6.3–13.3	4.7–10.2	8.3–14.5	
Overall removal efficiency	48%	92%	89%	50%	74%	41%	
Aerial tissue	6.98	0.01	0.16	0.17	—	1.33	[133]
Aerial tissue	288	0.05	—	0.09	—	—	[134]
Aerial tissue	—	4.4–13.5	—	36–108	—	75–143	[135]
Aerial tissue	7.0	—	12.7	—	—	20.3	[136]

Table 11.
Typical metal uptake rates and primary modes of removal using constructed wetlands.

3. Conclusions and future direction

One of the consequences of a rapidly growing global population is the increasing generation of municipal and industrial wastewaters, and leachates commensurate with corresponding increases in metal emissions. Metal-rich wastewaters can result in short- and long-term environmental damage with toxicity often demonstrated at low concentrations. Metal recovery from wastewaters is therefore becoming a significant issue not only because of its resulting environmental damage and associated health impacts but also because of its potential economic value.

In this chapter we review the effectiveness of biochar, microbial and lignin biosorbents as well as constructed wetland systems to remove soluble metals from wastewaters. There exists a wide variation in the adsorptive capacities of the various unmodified biosorbent materials reflecting the heterogeneity of the source materials used for their production. In a manner similar to biochars and lignins, metal removal using selected microbes yields a range of uptake levels, with high Pb removal rates using the strains *Paenibacillus peoriae strainTS7* and *Providencia vermicola strain SJ2A*. Physical (e.g. steam, high temperature) and chemical (e.g. acids/bases) modifications of biochars bring about improvements in metal uptake levels, and interestingly, modification of lignins produces noticeably higher metal adsorption rates with many of the metal uptake levels in the range 50–400 mg g^{-1} depending on the lignin source, particular type of modification and specific metal adsorbed. Modification of lignins leads to a significant increase in surface functionality by increasing the number of oxygen-, nitrogen- and sulphur-containing surface groups, while wood-based activated carbons also have a high affinity for many of the metals of interest, with uptake rates in many cases similar to those of modified lignins. While parameters such as contact time, adsorbent dosage, temperature and ionic strength play an important role in biosorption efficiency, the influence of solution chemistry and pH in particular also plays a highly significant role in the effective binding of a metal species to biosorbent materials. The solution pH also plays a key role in determining the oxidation state, ionic nature and solubility of the metal species. However the pH at which maximum adsorption occurs may also simultaneously precipitate metals from solution, and it is important that reported metal adsorption rates in any experimental work do not also (and erroneously) include precipitated metals. Many of the studies presented in this chapter have reported maximum metal biosorption rates at pH 4–7, but some are as low as pH 2 and others as high as pH 8.

Constructed wetlands differ from biosorbents in that they are biosystems which remove metals and other contaminants by sedimentation as well as physical and biochemical transformations. Many of the metals in the wastewater are removed by the macrophytic system which if harvested can be permanently removed without leaching back into the system by natural plant decay. The specific removal rates in CW systems are much lower than those of corresponding biosorbents; however with sufficient hydraulic retention time and appropriate loading rates, the overall removal efficiencies can be relatively high. Because of their passive nature, low-energy and low-maintenance requirements as well as their perceived amenity benefits, CWs are considered a promising technology for removal of metals from wastewater. There is however scope for further investigation into CWs treating metal-rich wastewaters such as (i) their long-term capacity to treat and retain the adsorbed metals from wastewaters which have a matrix of metals and (ii) assessing the influence of wastewater characteristics such as suspended solids, pH and predominance of metal types on CW removal efficiencies.

While application of an abundant supply of biosorbent materials to remove metals from wastewater is gaining increasing attention due to their potential for metal recovery and pollution mitigation, there are nevertheless a number of shortcomings to be addressed before their widespread use can be implemented. Some of the most commonly identified improvements are to increase the selectivity of biosorbents to treat wastewaters which have a matrix of metals and to improve their consistency, mechanical stability and adsorption capacity, making them less sensitive to pH changes and high ionic concentrations. In addition there is a need to develop the desorption potential and regeneration capacity of biosorbents in order to increase their technology readiness level. There is a need also to improve the long-term mechanical stability of biosorbents which generally deteriorates after a number of cycles. Possible ways to achieve this might be to immobilise the biosorbents onto inert materials such as sand, glass or fibres or perhaps use an entrapment technique in polymeric matrices producing alginate or polyacrylamide beads.

There are a range of potential biological materials that could be utilised and developed as strategies to remove metals from waste streams. Their further use will depend on developing engineering and technological solutions for their full deployment. Hand in hand with the removal of metals, there is also an interest in strategic deployment of such techniques to enhance specificity of metal binding; this interest stems from the need to recover metals of particular interest such as rare earth metals, valuable metals or radioisotopes.

Acknowledgements

The authors acknowledge support from the Geological Survey of Ireland (GSI, project no. 2018-ERAMIN2-002), the Irish Environmental Protection Agency (EPA) and an EU ERA-MIN2 award to the EU Biomimic Consortium (ID 86).

Conflict of interest

The authors declare no conflict of interest.

Author details

John G. Murnane[3]*, Bashir Ghanim[1,4], Lisa O'Donoghue[3], Ronan Courtney[2,4], Thomas F. O'Dwyer[1,4] and J. Tony Pembroke[1,4]

1 Department of Chemical Sciences, School of Natural Sciences, University of Limerick, Limerick, Ireland

2 Department of Biological Science, School of Natural Sciences, University of Limerick, Limerick, Ireland

3 School of Natural Sciences, School of Engineering, University of Limerick, Limerick, Ireland

4 Bernal Institute, University of Limerick, Limerick, Ireland

*Address all correspondence to: john.murnane@ul.ie

References

[1] Gunatilake SK. Methods of removing heavy metals from industrial wastewater. Journal of Multidisciplinary Engineering Science Studies. 2015;**1**:12-18

[2] Pembroke JT, Naveena B, Armshaw P. The potential of the photoautotroph synechocystis for metal bioremediation. In: Shiomi N, editor. Advances in Bioremediation of Wastewater and Polluted Soil. London, UK: IntechOpen; 2015. pp. 1-22. ISBN 978-953-51-4228-7

[3] Abramov S, He J, Wimmer D, Lemloh M-L, Muehe EM, Gann B, et al. Heavy metal mobility and valuable contents of processed municipal solid waste incineration residues from Southwestern Germany. Waste management. 2018;**79**:735-743. DOI: 10.1016/j.wasman.2018.08.010

[4] Dean JG, Bosqui FL, Lanouette KH. Removing heavy metals from waste water. Journal of Environmental Science and Technology. 1972;**6**:518-522. DOI: 10.1021/es60065a006

[5] European Drinking Water Directive (Council Directive 98/83/EC of 3 November 1998 on the Quality of Water Intended for Human Consumption). http:/ec.europa.eu/environment/water/water-drink/legislation_en.html [Accessed 2018.11.24]

[6] U.S. Safe Drinking Water Act (SDWA). https://www.epa.gov/sdwa [Accessed 2018.11.24]

[7] Australian Drinking Water Guidelines. 2011. https://nhmrc.gov.au/about-us/publications/australian-drinking-water-guidelines#block-views-block-file-attachments-content-block-1 [Accessed 2018.11.24]

[8] WHO Guidelines for Drinking Water Quality. 4th ed. https://www.who.int/water_sanitation_health/publications/drinking-water-quality-guidelines-4-including-1st-addendum/en/ [Accessed 2018.11.24]

[9] Charerntanyarak L. Heavy metals removal by chemical coagulation and precipitation. Water Science and Technology. 1999;**39**:135-138

[10] Li YJ, Zeng XP, Liu YF, Yan SS, Hu ZH, Ni Y. Study on the treatment of copper-electroplating wastewater by chemical trapping and flocculation. Separation and Purification Technology. 2003;**31**:91-95. DOI: 10.1016/S1383-5866(02)00162-4

[11] Lazaridis NK, Matis KA, Webb M. Flotation of metal-loaded clay anion exchangers. Part I: The case of chromate. Chemosphere. 2001;**42**:373-378. DOI: 10.1016/S0045-6535(00)00143-0

[12] Walton R. Zinc cementation. In: Developments in Mineral Processing. Vol. 15. Amsterdam, Netherlands: Elsevier; 2005. pp. 589-601 (Chapter 24)

[13] Holt PK, Barton GW, Mitchell CA. The future for electrocoagulation as a localized water treatment technology. Chemosphere. 2005;**59**:355-367. DOI: 10.1016/j.chemosphere.2004.10.023

[14] Pansini M, Colella C, De Gennaro M. Chromium removal from water by ion exchange using zeolite. Desalination. 1991;**83**:145-157. DOI: 10.1016/0011-9164(91)85091-8

[15] Abdel-Halim SH, Shehata AMA, El-Shahat MF. Removal of lead ions from industrial waste water by different types of natural materials. Water Research. 2003;**37**:1678-1683. DOI: 10.1016/S0043-1354(02)00554-7

[16] Juang RS, Shiau RC. Metal removal from aqueous solutions using chitosan-enhanced membrane filtration. Journal

of Membrane Science. 2000;**165**:159-167. DOI: 10.1016/S0376-7388(99)00235-5

[17] Rether A, Schuster M. Selective separation and recovery of heavy metal ions using water-soluble *N*-benzoylthiourea modified PAMAM polymers. Reactive and Functional Polymers. 2003;**57**:13-21. DOI: 10.1016/j.reactfunctpolym.2003.06.002

[18] Barakat MA, Schmidt E. Polymer-enhanced ultrafiltration process for heavy metals removal from industrial wastewater. Desalination. 2010;**256**: 90-93. DOI: 10.1016/j.desal.2010.02.008

[19] Woodward and Curran Inc Industrial Waste Treatment Handbook. 2nd ed. Burlington, MA, USA: Elsevier Butterworth Heinemann; 2006. ISBN 13:978-0-7506-7963-3

[20] Eccles H. Treatment of metal-contaminated wastes: Why select a biological process? Trends in Biotechnology. 1999;**17**:462-465. DOI: 10.1016/S0167-7799(99)01381-5

[21] Fu F, Wang Q. Removal of heavy metal ions from wastewaters: A review. Journal of Environmental Management. 2011;**92**:407-418. DOI: 10.1016/j.jenvman.2010.11.011

[22] Johnson PD, Girinathannair P, Ohlinger KN, Ritchie S, Teuber L, Kirby J. Enhanced removal of heavy metals in primary treatment using coagulation and flocculation. Water Environmental Research. 2008;**80**:472-479. DOI: 10.2175/106143007X221490

[23] López-Maldonado EA, Oropeza-Guzman MT, Jurado-Baizaval JL, Ochoa-Terán A. Coagulation–flocculation mechanisms in wastewater treatment plants through zeta potential measurements. Journal of Hazardous Materials. 2014;**279**:1-10. DOI: 10.1016/j.jhazmat.2014.06.025

[24] Tonini GA, Ruotolo LAM. Heavy metal removal from simulated wastewater using electrochemical technology: Optimization of copper electrodeposition in a membraneless fluidized bed electrode. Clean Technologies and Environmental Policy. 2017;**19**:403-415. DOI: 10.1007/s10098-016-1226-8

[25] Al-Enezi G, Hamoda MF, Fawzi N. Ion exchange extraction of heavy metals from wastewater sludges. Journal of Environmental Science and Health, Part A Toxic/Hazardous Substances and Environmental Engineering. 2011;**39**:455-464. DOI: 10.1081/ESE-120027536

[26] Li Y, Yang L, Xu Z, Sun Q. Separation and recovery of heavy metals from waste water using synergistic solvent extraction. IOP Conference Series: Materials Science and Engineering. 2017;**167**:012005. DOI: https://doi.org/10.1088/1757-899X/167/1/012005 [Accessed: 2018-11-24]

[27] Jeppesen T, Shu L, Keir G, Jegatheesan V. Metal recovery from reverse osmosis concentrate. Journal of Cleaner Production. 2009, 2009;**17**:703-707. DOI: 10.1016/j.jclepro.2008.11.013

[28] Baek K, Kim BK, Cho HJ, Yang JW. Removal characteristics of anionic metals by micellar-enhanced ultrafiltration. Journal of Hazardous Materials. 2003;**99**:303-311. DOI: 10.1016/S0304-3894(03)00063-3

[29] Leung WC, Wong MF, Chua H, Lo W, Leung CK. Removal and recovery of heavy metals by bacteria isolated from activated sludge treating industrial effluents and municipal wastewater. Water Science and Technology. 2000;**41**:233-240

[30] Kurniawan TA, Chan GYS, Lo WH, Babel S. Comparisons of low-cost adsorbents for treating wastewaters laden with heavy metals. Science of the Total Environment. 2005;**366**:409-426. DOI: 10.1016/j.scitotenv.2005.10.001

[31] Doyle FM, Liu ZD. The effect of triethylenetetraamine (trien) on the ion flotation of Cu^{2+} and Ni^{2+}. Journal of Colloid and Interface Science. 2003;**258**:396-403. DOI: 10.1016/S0021-9797(02)00092-9

[32] Allesina G, Pedrazzi S, Allegretti F, Morselli N, Puglia M, Santunione G, et al. Gasification of cotton crop residues for combined power and biochar production in Mozambique. Applied Thermal Engineering. 2018;**139**:387-394. DOI: 10.1016/j.applthermaleng.2018.04.115

[33] Luo L, Xu C, Chen Z, Zhang S. Properties of biomass-derived biochars: Combined effects of operating conditions and biomass types. Bioresource Technology. 2015;**192**:83-89. DOI: 10.1016/j.biortech.2015.05.054

[34] Wang S, Gao B, Zimmerman AR, Li Y, Ma L, Harris WG, et al. Physicochemical and sorptive properties of biochars derived from woody and herbaceous biomass. Chemosphere. 2015;**134**:257-262. DOI: 10.1016/j.chemosphere.2015.04.062

[35] Chen D, Liu D, Zhang H, Chen Y, Li Q. Bamboo pyrolysis using TG-FTIR and a lab-scale reactor: Analysis of pyrolysis behavior, product properties, and carbon and energy yields. Fuel. 2015;**148**:79-86. DOI: 10.1016/j.fuel.2015.01.092

[36] Chen X, Chen G, Chen L, Chen Y, Lehmann J, McBride MB, et al. Adsorption of copper and zinc by biochars produced from pyrolysis of hardwood and corn straw in aqueous solution. Bioresource Technology. 2011;**102**:8877-8884. DOI: 10.1016/j.biortech.2011.06.078

[37] Ghanim BM, Pandey DS, Kwapinski W, Leahy JJ. Hydrothermal carbonisation of poultry litter: Effects of treatment temperature and residence time on yields and chemical properties of hydrochars. Bioresource Technology. 2016;**216**:373-380. DOI: 10.1016/j.biortech.2016.05.087

[38] Wnetrzak R, Leahy JJ, Chojnacka KW, Saeid A, Novotny E, Jensen LS, et al. Influence of pig manure biochar mineral content on Cr(III) sorption capacity. Journal of Chemical Technology & Biotechnology. 2014;**89**:569-578. DOI: 10.1002/jctb.4159

[39] Lee Y, Eum P, Ryu C, Park Y, Jung J, Hyun S. Characteristics of biochar produced from slow pyrolysis of Geodae-Uksae 1. Bioresource Technology. 2013;**130**:345-350. DOI: 10.1016/j.biortech.2012.12.012

[40] Lima IM, Marshall WE. Adsorption of selected environmentally important metals by poultry manure-based granular activated carbons. Journal of Chemical Technolnology & Biotechnology. 2005;**80**(9):1054-1061. DOI: 10.1002/jctb.1283

[41] Lou K, Rajapaksha AU, Ok YS, Chang SX. Pyrolysis temperature and steam activation effects on sorption of phosphate on pine sawdust biochars in aqueous solutions. Chemical Speciation and Bioavailability. 2016;**28**:42-50. DOI: 10.1080/09542299.2016.1165080

[42] Jin H, Capareda S, Chang Z, Gao J, Xu Y, Zhang J. Biochar pyrolytically produced from municipal solid wastes for aqueous As (V) removal: Adsorption property and its improvement with KOH activation. Bioresource Technology. 2014;**169**:622-629. DOI: 10.1016/j.biortech.2014.06.103

[43] Xue Y, Gao B, Yao Y, Inyang M, Zhang M, Zimmerman AR, et al. Hydrogen peroxide modification enhances the ability of biochar (hydrochar) produced from hydrothermal carbonization of peanut hull to remove aqueous heavy metals: Batch and column tests. Chemical Engineering Journal.

2012;**200-202**:673-680. DOI: 10.1016/j.cej.2012.06.116

[44] Wang S, Gao B, Li Y, Mosa A, Zimmerman AR, Ma LQ, et al. Manganese oxide-modified biochars: Preparation, characterization, and sorption of arsenate and lead. Bioresource Technology. 2015;**181**:13-17. DOI: 10.1016/j.biortech.2015.01.044

[45] Samsuri AW, Sadegh-Zadeh F, Seh-Bardan BJ. Adsorption of As(III) and As(V) by Fe coated biochars and biochars produced from empty fruit bunch and rice husk. Journal of Environmental Chemical Engineering. 2013;**1**:981-988. DOI: 10.1016/j.jece.2013.08.009

[46] Pan J, Jiang J, Xu R. Adsorption of Cr(III) from acidic solutions by crop straw derived biochars. Journal of Environmental Sciences. 2013;**25**(10):1957-1965. DOI: 10.1016/S1001-0742(12)60305-2

[47] Dong X, Ma LQ, Li Y. Characteristics and mechanisms of hexavalent chromium removal by biochar from sugar beet tailing. Journal of Hazardous Materials. 2011;**190**:909-915. DOI: 10.1016/j.jhazmat.2011.04.008

[48] Xu D, Zhao Y, Sun K, Gao B, Wang Z, Jin J, et al. Cadmium adsorption on plant- and manure-derived biochar and biochar-amended sandy soils: Impact of bulk and surface properties. Chemosphere. 2014;**111**:320-326. DOI: 10.1016/j.chemosphere.2014.04.043

[49] Lu H, Zhang W, Yang Y, Huang X, Wang S, Qiu R. Relative distribution of Pb^{2+} sorption mechanisms by sludge-derived biochar. Water Research. 2012;**46**:854-862. DOI: 10.1016/j.watres.2011.11.058

[50] Kılıc M, Kırbıyık C, Cepeliogullar O, Pütüna AE. Adsorption of heavy metal ions from aqueous solutions by bio-char, a by-product of pyrolysis. Applied Surface Science. 2013;**283**:856-862. DOI: 10.1016/j.apsusc.2013.07.033

[51] Kołodynska D, Wnetrzak R, Leahy JJ, Hayes MHB, Kwapinski W, Hubicki Z. Kinetic and adsorptive characterization of biochar in metal ions removal. Chemical Engineering Journal. 2012;**197**:295-305. DOI: 10.1016/j.cej.2012.05.025

[52] Ma Y, Liu W, Zhang N, Li Y, Jiang H, Sheng G. Polyethylenimine modified biochar adsorbent for hexavalent chromium removal from the aqueous solution. Bioresource Technology. 2014;**169**:403-408. DOI: 10.1016/j.biortech.2014.07.014

[53] An Q, Jiang Y, Nan H, Yu Y, Jiang J. Unraveling sorption of nickel from aqueous solution by $KMnO_4$ and KOH-modified peanut shell biochar: Implicit mechanism. Chemosphere. 2019;**214**:846-854. DOI: 10.1016/j.chemosphere.2018.10.007

[54] Tan G, Sun W, Xu Y, Wang H, Xu N. Sorption of mercury (II) and atrazine by biochar, modified biochars and biochar based activated carbon in aqueous solution. Bioresource Technology. 2016;**211**:727-735. DOI: 10.1016/j.biortech.2016.03.147

[55] Tang J, Lv H, Gong Y, Huang Y. Preparation and characterization of a novel graphene/biochar composite for aqueous phenanthrene and mercury removal. Bioresource Technology. 2015;**196**:355-363. DOI: 10.1016/j.biortech.2015.07.047

[56] Wang S, Gao B, Li Y. Enhanced arsenic removal by biochar modified with nickel (Ni) and manganese (Mn) oxyhydroxides. Journal of Industrial and Engineering Chemistry. 2016;**37**:361-365. DOI: 10.1016/j.jiec.2016.03.048

[57] Jin J, Li S, Peng X, Liu W, Zhang C, Yang Y, et al. HNO_3 modified biochars for uranium (VI) removal from aqueous

solution. Bioresource Technology. 2018;**256**:247-253. DOI: 10.1016/j.biortech.2018.02.022

[58] Peng H, Gao P, Chu G, Pan B, Peng J, Xing B. Enhanced adsorption of Cu(II) and Cd(II) by phosphoric acid-modified biochars. Environmental Pollution. 2017;**229**:846-853. DOI: 10.1016/j.envpol.2017.07.004

[59] Bååth E. Effects of heavy metals in soil on microbial processes and populations (a review). Water, Air, and Soil Pollution. 1989;**47**:335-379. DOI: 10.1007/BF00279331

[60] Roane TM, Kellogg ST. Characterization of bacterial communities in heavy metal contaminated soils. Canadian Journal of Microbiology. 1996;**42**:593-603. DOI: 10.1139/m96-080

[61] Lovley DR, Coates JD. Bioremediation of metal contamination. Current Opinion in Biotechnology. 1997;**8**:285-289. DOI: 10.1016/S0958-1669(97)80005-5

[62] Bang SW, Clark DS, Keasling JD. Engineering hydrogen sulfide production and cadmium removal by expression of the thiosulfate reductase gene (*phsABC*) from *Salmonella enterica* serovar typhimurium in *Escherichia coli*. Applied and Environmental Microbiology. 2000;**66**:3939-3944. DOI: 10.1128/AEM.66.9.3939-3944.2000

[63] Sar P, Kazy SK, Singh SP. Intracellular nickel accumulation by *Pseudomonas aeruginosa* and its chemical nature. Letters in Applied Microbiology. 2001;**32**:257-261. DOI: 10.1046/j.1472-765X.2001.00878.x

[64] Schembri MA, Kjaergaard K, Klemm P. Bioaccumulation of heavy metals by fimbrial designer adhesions. FEMS Microbiology Letters. 1999;**170**:363-371. DOI: 10.1111/j.1574-6968.1999.tb13396.x

[65] White C, Sharman AK, Gadd GM. An integrated microbial process for the bioremediation of soil contaminated with toxic metals. Nature Biotechnology. 1998;**16**:572-575

[66] Vakilchap F, Mousavi SM, Shojaosadati SA. Role of *Aspergillus niger* in recovery enhancement of valuable metals from produced red mud in Bayer process. BioresourceTechnology. 2016;**218**:991-998. DOI: 10.1016/j.biortech.2016.07.059

[67] Silver S. Exploiting heavy-metal resistance systems in bioremediation. Research in Microbiology. 1994;**145**:61-67. DOI: 10.1016/0923-2508(94)90072-8

[68] Nies DH. Microbial heavy-metal resistance. Applied Microbiology and Biotechnology. 1999;**51**:730-750. DOI: 10.1007/s002530051457

[69] Bruins MR, Kapil S, Oehme FW. Microbial resistance to metals in the environment. Ecotoxicology and Environmental Safety. 2000;**45**:198-207. DOI: 10.1006/eesa.1999.1860

[70] Pembroke JT, Piterina AV. A novel ICE in the genome of Shewanella putrefaciens W3-18-1: Comparison with the SXT/R391 ICE-like elements. FEMS Microbiology Letters. 2006;**264**:80-88. DOI: 10.1111/j.1574-6968.2006.00452.x

[71] Bazaka K, Crawford RJ, Nazarenko EL, Ivanova EP. Bacterial extracellular polysaccharides. In: Linke D, Coldman A, editors. Bacterial Adhesion. Adv. Exp. Med. Biol. Dordrecht, Netherlands: Springer; 2011. pp. 213-223. DOI: 10.1007/978-94-007-0940-9_13

[72] Gadd GM. Biosorption: Critical review of scientific rationale, environmental importance and significance for pollution treatment. Journal of Chemical Technology and Biotechnology. 2009;**84**:13-28. DOI: 10.1002/jctb.1999

[73] Pirszel J, Pawlik B, Skowronski T. Cation-exchange capacity of algae and

cyanobacteria: A parameter of their metal sorption abilities. Journal of Industrial Microbiology. 1995;**14**:319-322. DOI: 10.1007/BF01569945

[74] Pan X, Liu J, Song W, Zhang D. Biosorption of Cu(II) to extracellular polymeric substances (EPS) from *Synechocystis sp*: A fluorescence quenching study. Frontiers of Environmental Science and Engineering. 2012;**6**:493-497. DOI: 10.1007/s11783-012-0416-9

[75] Ozturk S, Aslimb B, Suludereb Z, Tan S. Metal removal of cyanobacterial exopolysaccharides by uronic acid content and monosaccharide composition. Carbohydrate Polymers. 2014;**101**:265-271. DOI: 10.1016/j.carbpol.2013.09.040

[76] Kumara R, Singha K, Sarkarb S, Sethib LN. Accumulation of Cu by microalgae Scenedesmus obliquus and Synechocystis sp. PCC 6803. IOSR Journal of Environmental Science, Toxicology and Food Technology. 2014;**8**:64-68

[77] Costley SC, Wallis FM. Bioremediation of heavy metals in a synthetic wastewater using a rotating biological contactor. Water Research. 2001;**35**:3715-3723. DOI: 10.1016/j.carbpol.2013.09.040/10.1016/S0043-1354(01)00072-0

[78] Sasaki K, Morikawa H, Kishibe T, Mikami A, Harada T, Ohta M. Practical removal of radioactivity from sediment mud in a swimming pool in Fukushima, Japan by immobilized photosynthetic bacteria. Bioscience, Biotechnology and Biochemistry. 2012;**76**:859-866. DOI: 10.1271/bbb.110853

[79] Fella-Temzia S, Yalaoui-Guellala D, Rodriguez-Carvajalb MA, Belhadic D, Madania K, Kaci Y. Removal of lead by exopolysaccharides from Paenibacillus peoriae strainTS7 isolated from rhizosphere of durum wheat. Biocatalysis and Agricultural Biotechnology. 2018;**16**:425-432. DOI: 10.1016/j.bcab.2018.09.016

[80] Goswami S, Syiem MB, Pakshirajan K. Cadmium removal by *Anabaena doliolum* Ind1 isolated from a coal mining area in Meghalaya, India: Associated structural and physiological alterations. Environmental Engineering Research. 2015;**20**(1):41-50. DOI: 10.4491/eer.2014.059

[81] Tassist A, Lounici H, Abdi N, Mameri N. Equilibrium, kinetic and thermodynamic studies on aluminum biosorption by a mycelial biomass (*Streptomyces rimosus*). Journal of Hazardous Materials. 2010;**183**:35-43. DOI: 10.1016/j.jhazmat.2010.06.078

[82] Sharma J, Shamim K, Dubey SK, Meena RM. Metallothionein assisted periplasmic lead sequestration as lead sulfite by Providencia vermicola strain SJ2A. Science of the Total Environment. 2017;**579**:359-365. DOI: 10.1016/j.scitotenv.2016.11.089

[83] Deepika K, Raghuram M, Kariali E, Bramhachari P. Biological responses of symbiotic *Rhizobium radiobacter* strain VBCK1062 to the arsenic contaminated rhizosphere soils of mung bean. Ecotoxicology and Environmental Safety. 2016;**134**:1-10. DOI: 10.1016/j.ecoenv.2016.08.008

[84] Cayllahua JEB, Torem ML. Biosorption of aluminum ions onto Rhodococcus opacus from wastewaters. Chemical Engineering Journal. 2010;**161**:1-8. DOI: 10.1016/j.cej.2010.03.025

[85] Gupta P, Diwan B. Bacterial exopolysaccharide mediated heavy metal removal: A review on biosynthesis, mechanism and remediation strategies. Biotechnology Reports. 2017;**13**:58-71. DOI: 10.1016/j.btre.2016.12.006

[86] Vanholme R, Morreel K, Ralph J, Boerjan W. Lignin engineering.

Current Opinion in Plant Biology. 2008;**11**(3):278-285. DOI: 10.1016/j.pbi.2008.03.005

[87] Thakur VK, Thakur MK, Raghavan P, Kessler MR. Progress in green polymer composites from lignin for multifunctional applications: A review. ACS Sustainable Chemistry & Engineering. 2014;**2**:1072-1092. DOI: 10.1021/sc500087z

[88] Wu Y, Zhang S, Guo X, Huang H. Adsorption of chromium(III) on lignin. Bioresource Technology. 2008;**99**(16):7709-7715. DOI: 10.1016/j.biortech.2008.01.069

[89] Mohan D, Pittman CU Jr, Steele PH. Single, binary and multi-component adsorption of copper and cadmium from aqueous solutions on Kraft lignin—A biosorbent. Journal of Colloid and Interface Science. 2006;**297**(2):489-504. DOI: 10.1016/j.jcis.2005.11.023

[90] Merdy P, Guillon E, Aplincourt M, Dumonceau J, Vezin H. Copper sorption on a straw lignin: Experiments and EPR characterization. Journal of Colloid and Interface Science. 2002;**245**(1):24-31. DOI: 10.1006/jcis.2001.7972

[91] Todorciuc T, Bulgariu L, Popa VI. Adsorption of Cu(II) from aqueous solution on wheat straw lignin: Equilibrium and kinetic studies. Cellulose Chemistry and Technology. 2015;**49**:439-447

[92] Demirbas A. Adsorption of lead and cadmium ions in aqueous solutions onto modified lignin from alkali glycerol delignication. Journal of Hazardous Materials. 2004;**109**:221-226. DOI: 10.1016/j.jhazmat.2004.04.002

[93] Srivastava SK, Singh AK, Sharma A. Studies on the uptake of lead and zinc by lignin obtained from black liquor—A paper industry waste material. Environmental Technology. 1994;**15**(4):353-361. DOI: 10.1080/09593339409385438

[94] Ge Y, Li Z. Application of lignin and its derivatives in adsorption of heavy metal ions in water: A review. ACS Sustainable Chemistry & Engineering. 2018;**6**:7181-7192

[95] Dizhbite T, Jashina L, Dobele G, Andersone A, Evtuguin D, Bikovens O, et al. Polyoxometalate (POM)-aided modification of lignin from wheat straw biorefinery. Holzforschung. 2013;**67**:539-547. DOI: 10.1515/hf-2012-0193

[96] Quintana GC, Rocha GJM, Goncalves AR, Velasquez JA. Evaluation of heavy metal removal by oxidised lignins in acid media from various sources. BioResources. 2008;**3**:1092-1102

[97] Supanchaiyamat N, Jetsrisuparb K, Knijnenburg JTN, Tsang DCW, Hunt AJ. Lignin materials for adsorption: Current trend, perspectives and opportunities. Bioresource Technology. 2019;**272**:570-581. (in press)

[98] Ge Y, Song Q, Li Z. A Mannich base biosorbent derived from alkaline lignin for lead removal from aqueous solution. Journal of Industrial and Engineering Chemistry. 2015;**23**:228-234. DOI: 10.1016/j.jiec.2014.08.021

[99] Tian J, Ren S, Fang G, Ma Y, Ai Q. Preparation and performance of dimethyl-acetoxy-(2-carboxymethyl ether)-lignin ammonium chloride amphoteric surfactant. BioResources. 2014;**9**:6290-9303

[100] Peternele WS, Winkler-Hechenleitner AA, Gómez Pineda EA. Adsorption of Cd(II) and Pb(II) onto functionalized formic lignin from sugar cane bagasse. Bioresource Technology. 1999;**68**:95-100

[101] Liang F-B, Song Y-L, Huang C-P, Li Y-X, Chen B-H. Synthesis of novel

lignin-based ion-exchange resin and its utilization in heavy metals removal. Industrial & Engineering Chemistry Research. 2013;**52**:1267-1274. DOI: 10.1021/ie301863e

[102] Parajuli D, Inoue K, Ohto K, Oshima T, Murota A, Funaoka M, et al. Adsorption of heavy metals on crosslinked lignocatechol: A modified lignin gel. Reactive and Functional Polymers. 2004;**62**:129-139. DOI: 10.1016/j.reactfunctpolym.2004.11.003

[103] Liu XL, Zhu HX, Qin CR, Zhou JH, Zhao JR, Wang SF. Adsorption of heavy metal ion from aqueous single metal solution by animated epoxy-lignin. BioResources. 2013;**8**:2257-2269

[104] Li Z, Kong Y, Ge Y. Synthesis of porous lignin xanthate resin for Pb^{2+} removal from aqueous solution. Chemical Engineering Journal. 2015;**270**:229-234. DOI: 10.1016/j.cej.2015.01.123

[105] Ge Y, Xiao D, Li Z, Cui X. Dithiocarbamate functionalized lignin for efficient removal of metallic ions and the usage of the metal-loaded bio-sorbents as potential free radical scavengers. Journal of Materials Chemistry A. 2014;**2**:2136-2145. DOI: 10.1039/c3ta14333c

[106] Jin C, Zhang X, Xin J, Liu G, Wu G, Kong Z, et al. Clickable synthesis of 1,2,4-Triazole modified lignin-based adsorbent for the selective removal of Cd(II). ACS Sustainable Chemistry & Engineering. 2017;**5**:4086-4093. DOI: 10.1021/acssuschemeng.7b00072

[107] Li Z, Xiao D, Ge Y, Koehler S. Surface-functionalized porous lignin for fast and efficient lead removal from aqueous solution. ACS Applied Materials & Interfaces. 2015;**7**:15000-15009. DOI: 10.1021/acsami.5b03994

[108] Suhas, PJM C, Carrott MML. Lignin—From natural adsorbent to activated carbon: A review. Bioresource Technology. 2007;**98**:2301-2312. DOI: 10.1016/j.biortech.2006.08.008

[109] Danish M, Ahmad T. A review on utilization of wood biomass as a sustainable precursor for activated carbon production and application. Renewable and Sustainable Energy Reviews. 2018;**87**:1-21. DOI: 10.1016/j.rser.2018.02.003

[110] Chistyakov AV, Tsodikov MV. Methods for preparing carbon sorbents from lignin (review). Russian Journal of Applied Chemistry. 2018;**91**:1090-1105. DOI: 10.1134/S1070427218070054

[111] Kalavathy MH, Karthikeyan T, Rajgopal S, Miranda LR. Kinetic and isotherm studies of Cu(II) adsorption onto H_3PO_4-activated rubber wood sawdust. Journal of Colloid and Interface Science. 2005;**292**:354-362. DOI: 10.1016/j.jcis.2005.05.087

[112] Kalavathy MH, Miranda LR. *Moringa oleifera*—A solid phase extractant for the removal of copper, nickel and zinc from aqueous solutions. Chemical Engineering Journal. 2010;**158**:188-199. DOI: 10.1016/j.cej.2009.12.039

[113] Sartape AS, Mandhare AM, Salvi PP, Pawar DK, Kolekar SS. Kinetic and equilibrium studies of the adsorption of Cd(II) from aqueous solutions by wood apple shell activated carbon. Desalination and Water Treatment. 2013;**51**:4638-4650. DOI: 10.1080/19443994.2012.759158

[114] Karthikeyan T, Rajgopal S, Miranda LR. Chromium(VI) adsorption from aqueous solution by *Hevea Brasilinesis* sawdust activated carbon. Journal of Hazardous Materials. 2005;**124**:192-199. DOI: 10.1016/j.jhazmat.2005.05.003

[115] Khezami L, Capart R. Removal of chromium(VI) from aqueous

solution by activated carbons: Kinetic and equilibrium studies. Journal of Hazardous Materials. 2005;**123**:223-231. DOI: 10.1016/j.jhazmat.2005.04.012

[116] Danish M, Hashim R, Ibrahim MNM, Rafatullah M, Sulaiman O. Surface characterization and comparative adsorption properties of Cr(VI) on pyrolysed adsorbents of *Acacia mangium* wood and *Phoenix dactylifera* L. stone carbon. Journal of Analytical and Applied Pyrolysis. 2012;**97**:19-28. DOI: 10.1016/j.jaap.2012.06.001

[117] Singh CK, Sahu JN, Mahalik KK, Mohanty CR, Mohan BR, Meikap BC. Studies on the removal of Pb(II) from wastewater by activated carbon developed from *Tamarind wood* activated with sulphuric acid. Journal of Hazardous Materials. 2008;**153**:221-228. DOI: 10.1016/j.jhazmat.2007.08.043

[118] Acharya J, Sahu JN, Mohanty CR, Meikap BC. Removal of lead(II) from wastewater by activated carbon developed from *Tamarind wood* by zinc chloride activation. Chemical Engineering Journal. 2009;**149**:249-262. DOI: 10.1016/j.cej.2008.10.029

[119] Mayes WM, Aumonier J, Jarvis AP. Preliminary evaluation of a constructed wetland for treating extremely alkaline (pH 12) steel slag drainage. Water Science and Technology. 2009;**59**:2253-2263

[120] Vymazal J. Constructed wetlands for wastewater treatment: Five decades of experience. Environmental Science & Technology. 2011;**45**:61-69. DOI: 10.1021/es101403q

[121] Kotti IP, Gikas GD, Tsihrintzis VA. Effect of operational and design parameters on removal efficiency of pilot-scale FWS constructed wetlands and comparison with HSF systems. Ecological Engineering. 2010;**36**:862-875. DOI: 10.1016/j.ecoleng.2010.03.002

[122] Kadlec RH, Wallace S. Treatment Wetlands. CRC Press; 2008

[123] Stottmeister U, Wießner A, Kuschk P, Kappelmeyer U, Kästner M, Bederski O, et al. Effects of plants and microorganisms in constructed wetlands for wastewater treatment. Biotechnology Advances. 2003;**22**:93-117. DOI: 10.1016/j.biotechadv.2003.08.010

[124] Shelef O, Gross A, Rachmilevitch S. Role of plants in a constructed wetland: Current and new perspectives. Water. 2013;**5**:405-419. DOI: 10.3390/w5020405

[125] PIRAMID Consortium. Engineering Guidelines for the Passive Remediation of Acidic and/or Metalliferous Mine Drainage and Similar Wastewaters. Newcastle Upon Tyne: University of Newcastle; 2003

[126] Kosolapov DB, Kuschk P, Vainshtein MB, Vatsourina AV, Wiessner A, Kästner M, et al. Microbial processes of heavy metal removal from carbon-deficient effluents in constructed wetlands. Engineering in Life Sciences. 2004;**4**:403-411. DOI: 10.1002/elsc.200420048

[127] Johnson DB, Hallberg KB. Acid mine drainage remediation options: A review. Science of the Total Environment. 2005;**38**:3-14. DOI: 10.1016/j.scitotenv.2004.09.002

[128] Higgins D, Curtin T, Courtney R. Effectiveness of a constructed wetland for treating alkaline bauxite residue leachate: A 1-year field study. Environmental Science and Pollution Research. 2017;**24**:8516-8524. DOI: 10.1007/s11356-017-8544-1

[129] Vymazal J, Krása P. Distribution of Mn, Al, Cu and Zn in a constructed wetland receiving municipal sewage. Water Science and Technology. 2003;**48**:299-305

[130] Lesage E, Rousseau DPL, Meers E, Tack FMG, De Pauw N. Accumulation of metals in a horizontal subsurface flow constructed wetland treating domestic wastewater in Flanders, Belgium. Science of the Total Environment. 2007;**380**:102-115. DOI: 10.1016/j.scitotenv.2006.10.055

[131] Kröpfelová L, Vymazal J, Švehla J, Štíchová J. Removal of trace elements in three horizontal sub-surface flow constructed wetlands in the Czech Republic. Environ Pollution. 2009;**157**:1186-1194. DOI: 10.1016/j.envpol.2008.12.003

[132] Khan S, Ahmad MT, Shah S, Rehman A, Khaliq A. Use of constructed wetland for the removal of heavy metals from industrial wastewater. Journal of Environmental Management. 2009;**90**:3451-3457. DOI: 10.1016/j.jenvman.2009.05.026

[133] Vymazal J, Dusek J, Kvet J. Nutrient uptake and storage by plants in constructed wetlands with horizontal sub-surface flow: A comparative study. In: Vymazal J, editor. Nutrient Cycling and Retention in Natural and Constructed Wetlands. Leiden, The Netherlands: Backhuys Publishers; 1999. pp. 85-100

[134] Peverlya JH, Surface JM, Wang T. Growth and trace metal absorption by Phragmites australis in wetlands constructed for landfill leachate treatment. Ecological Engineering. 1995;**5**:21-35. DOI: 10.1016/0925-8574(95)00018-E

[135] Scholes LNL, Schutes RBE, Revitt DM, Purchase D, Forshaw M. The removal of urban pollutants by constructed wetlands during wet weather. Water Science & Technology. 1999;**40**:333-340

[136] Liu J-G, Li G-H, Shao W-C, Xu J-K, Wang D-K. Variations in uptake and translocation of copper, chromium and nickel among nineteen wetland plant species. Pedosphere. 2010;**20**:96-103. DOI: 10.1016/S1002-0160(09)60288-5

Chapter 6

Removal of *Escherichia Coli* Using Low-Frequency Electromagnetic Field in Riverbank Filtration

Rossitah Selamat, Ismail Abustan, Mohd Rizal Arshad and Nurul Hana Mokhtar Kamal

Abstract

An increase of pathogenic bacteria (*E. coli*) in river water is a concern as it is the main precursor to health hazard disinfection in conventional drinking water treatment systems. Riverbank filtration (RBF) is a non-chemical techniques and natural treatments that efficient in reducing or removing the contaminants in the water. Therefore, this study aimed to remove *Escherichia coli* (*E. coli*), and reduce the concentration with low-frequency electromagnetic fields (LF-EMF) as a component of the non-ionising radiations in RBF. This research design and construct a LF-EMF device on horizontal coiled columns that were capable of producing uniform magnetic fields in the frequency range of 50 Hz. A magnetic field density was varied at 2, 4, 6, 8, and 10 mT. The diameter of column was 50 mm, which underwent 6 hours of LF-EMF exposure at 50 mL/min of water flowrates. The maximum removal efficiency of *E. coli* in was 100% at 6, 8, and 10 mT of magnetic field exposure. These results indicated that the *E. coli* in the sample of water that was exposed to the LF-EMF was statistically significantly decreased. The magnetic intensity of the LF-EMF changed the characteristic responses for *E. coli* bacteria.

Keywords: electromagnetic field, low-frequency, *E. coli*., growth, river water

1. Introduction

Water is a fundamental need, and the most abundant of resources [1]. However, the World Health Organisation (WHO) stated that, in 2012, about 780 million people were without an adequate drinking water source [2]. Hence, the demand for good quality and clean drinking water has increased, especially among Malaysian consumers. Raw water originating from surface water and groundwater needs to be treated before the water is made potable. According to statistics in 2017, in Malaysia, 500 water treatment plants (WTP) were in operation to treat raw water, and produced about 16,536 million litres per day (MLD) of drinking water to consumers.

Clean and safe water is one of the most pressing global health-affecting and environmental issues. Generally, in Malaysia, surface water is exposed to organic, inorganic, and microbial pathogen contamination due to poor management of septic tanks, wastewater, and agricultural runoff and earthwork products [3]. Approximately 99% of water supply for domestic uses in Malaysia originate from surface water such as rivers and streams, while another 1% originate from groundwater [4]. The surface water in the country has also been polluted with, for example, biological contaminants such as viruses, bacteria, and protozoa which are capable of causing illnesses in humans like bloody diarrhoea, affecting human health as well as the environment [5].

According to the Department of Environment (DOE) Malaysia's annual report, 48% of the 473 rivers monitored in 2014 have been contaminated by these sources. This high percentage reflects that the water resources in Malaysia are contaminated, and the condition may continue to worsen. Among all the pollutant loads entering surface water, bio-colloids are the major pollutants attributed by wastewater discharge and surface runoff. These bio-colloids usually refer to microorganisms in the water, such as bacteria and protozoa [6]. About 842,000 death cases involving diarrhoeal illnesses because of drinking water contamination were reported [7]. The situation can worsen during extreme weathers such as El Nino (drought), and El Nina (floods) that have a great impact on the quality and quantity of the water resource [8]. According to previous studies, contamination of bacteria significantly increases in surface water during these events [9]. This poses more challenges to the authorities in delivering and providing safe drinking water via the conventional treatment system because of the low surface water level, and high pollutant loads [10]. Therefore, to ensure a stable and safe drinking water supply, alternative methods for water management are necessary especially during extreme weather conditions.

Riverbank filtering (RBF) is an attractive option that can be applied for effective water treatment. RBF is a technique that covers both shallow groundwater and river water that have crossed through the banks of rivers, or riverbanks to well extractions [11]. Most of the suspended and dissolved contaminants, including viruses and pathogenic bacteria, are filtered out as surface water is filtered through aquifer materials, and the sediments of the riverbed [12]. Abstracting of riverbank water can overcome water shortage due to extreme events such as floods and droughts that cause water levels to increase on the ground, or reduce underwater intake pipes, causing disruptions in water transfer to treatment plants [13]. Although RBF is a capable method for improving surface water quality, it does not abolish the problem. Abstracted well water quality is highly dependent on several factors, such as groundwater and river water quality, temperature and pH of water, water residence time, medium porosity, and oxygen concentrations [14]. According to Levantesi et al. study, the breakthrough of bacteria and turbidity occurred in a shallow drilled well (3–6 m) due to the short travel time, especially during monsoon seasons [15]. This condition urges for appropriate treatment applications to further enhance the ability of RBF in bacteria and inorganic substance removal.

Indicator bacteria, including the total coliforms, *Escherichia coli* (*E. coli*), *Enterococci*, and *Clostridium perfringens*, are commonly used to measure drinking and raw water quality. The presence of faecal coliform and *E. coli* is likewise a potable water contamination indicator through animal or human faecal matter [16]. *E. coli* bacteria indicate the potential presence of pathogenic microorganisms in natural and treated waters. *E. coli* can cause a variety of intestinal and extra-intestinal infections, such as diarrhoea, urinary tract infection, meningitis, peritonitis, septicemia, and gram-negative bacterial pneumonia [17]. To date, many treatment methods for the removal of *E. coli* have been introduced in treatment plants, such as membrane filtration [18], soil aquifer treatment [15], slow sand filtration [19], granular activated carbon (GAC) adsorption [20], and advanced oxidation [21]. All these methods have long been used in water treatment, and proved effective for bacteria removal. However, there is no information about non-ionising radiation applications in water treatment plants in Malaysia. This method is a better option for new applications in RBF systems based on the requirements of packing materials around the well screen.

Despite the numerous advantages of RBF, it also has several limitations [22]. Seasonal variations have an impact on the concentrations of *E. coli* in riverbeds, where they increased during the wet seasons [3]. Therefore, researchers have been conducting studies to explore and develop an efficient but cost-effective method capable of removing the *E. coli* using new application techniques. Low-frequency electromagnetic fields (LF-EMF) are a component of the non-ionising radiations used to treat and control the effective growth of *E. coli* bacteria [23]. Application of the LF-EMF on the *E. coli* bacteria has shown that exposure to non-ionising, electromagnetic radiation can induce numerous and quite varied removal effects [24]. Due to the capability of LF-EMF to remove the *E. coli* bacteria, this application was introduced as an alternative technique of *E. coli* removal in RBF. Therefore, evaluating the proposed LF-EMF effects on the RBF system is important to determine its effectiveness for the removal of *E. coli* in drinking water supply.

1.1 Malaysia's drinking water resources

Malaysia has had abundant and rich water resources throughout the years. The main source of drinking water in Malaysia is groundwater and surface water. Approximately 99% of water for domestic uses in Malaysia are from surface water, while another 1% of the supply is from groundwater [25]. Malaysia's internal water sources are estimated to be about 580 km^3/year, with 30% of water production for municipal uses [26]. Water supply from surface water is widely used as drinking water, such as water withdrawn from Sungai Kinta, Sungai Langat, and Sungai Selangor [27]. Water supply from groundwater intake from a few states in Malaysia such as Terengganu, Kelantan, Perlis, Kedah, Pahang, Sabah, and Sarawak are also used for drinking water [28]. According to the data published by Suruhanjaya Perkhidmatan Air Negara (SPAN) in 2015, only 1.5% of total groundwater supply is present in Malaysia, and there was an increase in groundwater usage by 3.3% from the year 2014 to 2015 (SPAN, 2015). Nonetheless, the key issue to be considered is the quality of the water sources for drinking water supply. Both surface and groundwater sources are easily affected by the surrounding changes, whether manmade or natural. Therefore, it is important to determine undesired constituents, and monitor the characteristics of the water sources to ensure the pollutants in the water do not exceed the standard limits for water supply stated by the National Water Quality Standards for Malaysia (NWQS), and the Ministry of Health (MOH).

1.1.1 Quality of water

The Department of Environment (DOE) uses the National Water Quality Standards for Malaysia (NWQS) and Water Quality Index (WQI) to evaluate the status of the water source quality [26]. The WQI, introduced by the DOE, has been practiced in Malaysia for about 25 years, and serves as the basis for the assessment of environment water quality, while the NWQS classifies the beneficial uses of the watercourse based on WQI [29]. To design the drinking water quality management system, the assessment of water quality is an important step in determining the possible problems in the quality of the drinking water source. Basically, the characteristics of water quality are determined by physical, chemical, and biological factors to describe the overall condition of the water quality and its suitability for a specific use.

1.1.2 Microorganism pollution

The occurrences of pollution and indicator pathogenic bacteria in potable water depend on a number of factors, including the intrinsic and chemical characteristics of the catchment area, and the range of human activities and animal sources that release pathogenic bacteria to the environment. Sources of pathogenic bacteria in potable water are numerous and, for operational efficiency, are typically assessed by faecal indicator bacteria investigation. In terms of biological characteristics for safe drinking water supply and drinking water distribution systems, water is one of the transmission routes for pathogenic microorganisms [2]. In spite of having enhanced water management and sanitation, waterborne-diseases and outbreaks may continue to occur [30]. Drinking water polluted by microorganisms of faecal origin is a current worldwide health concern because of epidemic occurrences globally in relation to microbial-contaminated water. In drinking water, these microorganisms of interest include protozoa, bacteria, viruses, algae, and helminths. An overview of these microorganisms is given in **Table 1**.

Faecal coliforms are bacteria which fulfil all the criteria used to define total coliforms, with the additional requirement that they grow and ferment lactose with the production of acid at a scientifically accurate 44.5°C [31]. This bacteria of the coliform subgroup has been found to have a positive correlation with faecal contamination of warm-blooded animals [32]. However, several thermotolerant coliform bacteria, by definition by the genus *Klebsiella* bacteria, have been isolated from environmental samples with the apparent absence of faecal pollution [33]. Similarly, Revetta et al. reported that other members of the thermotolerant coliform

Types	Description	Remarks
Bacteria *Vibrio Cholerae* *Escherichia coli* Legionella *Shigella* spp. *Samonella* spp.	• Single cell organism with size ranging from 0.1 to 10 μm. • Negatively charge surface • Aerobic, anaerobic, facultative • Motile and non-motile	• The most reported water-borne plaque are involve of bacteria
Protozoa *Cryptosporidium parvum Giardia lamblia* Entamoeba dispar *Entamoeba histolytica*	• Group of unicellular and non-photosynthetic organism with diameter size between 1 and 102 μm. • Negatively charge surface aerobic and anaerobic motile and non-motile	• Under water-borne disease stand-points, the four listed Protozoa are consider as the greatest risk in water supply
Virus T-4 coliphage Adenovirus Enterovirus Rotavirus MS-2 coliphage	• Smallest of waterborne agents with diameter size of 0.02–0.2 μm Negatively charge surface	• Poliovirus and Hepatitis A are the only known virus that have been documented to be associated with water-borne transmission
Algae Volvox Euglena Cyclotella Synedra Chlorella Anabaena	• Diameter size: 1–102 μm Negatively charge surface aerobic motile and non-motile	• Algae are common living organism in water supply and play important part in nutrient cycle. But a few algae are pathogenic to human because it produce endotoxins that can cause gastroenteritis
Helminths	• Diameter size: 1–102 μm Negatively charge surface aerobic motile	• Effective treatment and disposal of sewage water can control the parasitic worm in water supply

Table 1.
Microorganism in drinking water sources.

group and *E. coli* have been detected in clean areas, and were associated with regrowth events in the water distribution systems [34]. Faecal coliforms demonstrate a survival of the bacteria form similar to pathogenic bacteria, and also have usefulness as indicators of bacteria, tended to be replaced by *E. coli*.

Recently, the faecal coliform group has been extended to include other characteristics, such as β-D-*galactosidase*-positive reactions [35]. *E. coli* is a specific indicator for the presence of the faecal coliform group, and is the most reliable indicator of enteric pathogens [36]. Several studies have indicated that *E. coli* is an indicator of choice to indicate the occurrence of recent faecal coliform in drinking water. Currently, *E. coli* appears to provide the best bacterial indication of faecal coliform, and only several strains of *E. coli* in drinking water can cause diseases [37]. In several countries, this organism has been included in their regulations as a primary indicator of faecal contamination in drinking water [38]. Therefore, *E. coli* is the best faecal indicator to inform public-health risks associated with the consumption of contaminated drinking water.

Escherichia coli or *E. coli* is also known as a facultative anaerobic bacterium that is gram-negative. Cells of *E. coli* are typically rod-shaped with a cell volume of 0.6–0.7 μm^3, 2 μm long, and 0.5 μm in diameter [39]. Generally, *E. coli* is found in the faeces of healthy cattle, and is transmitted in the lower intestinal tracts of warm-blooded organisms, including humans and animals [32]. In 1885, this microorganism was discovered by Theodor Escherich, and was first classified as a human pathogen in 1982. Most of the *E. coli* strains harmlessly colonise the gastrointestinal tracts of humans and animals as normal flora. However, other strains grow into pathogenic *E. coli* by acquiring virulence, which is caused by bacteriophages, plasmids, transposons, and pathogenicity islands. Differences in survivability, external structure, size, shape, and zeta potential are some of the factors that influence the behaviour of these bacteria. These pathogenic *E. coli* can be categorised based on pathogenicity mechanisms, serotypes, clinical symptoms, or virulence factors. Several of the enterohaemorrhagic *E. coli* are defined as pathogenic *bacteria* that produce Shiga toxins, and cause the life-threatening sequelae of haemolytic uraemic syndrome, and haemorrhagic colitis in humans.

1.2 Riverbank filtration (RBF)

Subsurface or groundwater in Malaysia are natural water sources that can be exploited to meet the demands for water of high quality. The RBF process is an existing method referring to the process of extracting potable water at the riverbank, utilising subsurface or groundwater to supply sources of high quality water [28]. RBF systems and natural treatment processes typically take place during water infiltration. **Figure 1** shows the natural process of extracting treated water from an adjacent pumping well to a river.

As illustrated in the figure above, the difference in hydraulic gradient causes the water from the river to flow towards the well during the pumping process. Additionally, the RBF process is known as a sustainable and economical method to improve poor surface water quality. A complex attenuation method occurs during the transportation of water through the aquifer layer, resulting in raw water of high quality. The high quality raw water is then supplied to the water treatment plants, making it easier to be treated at low operating costs by conventional treatment systems. Therefore, water from the well can be directly consumed with very minimum treatment in certain areas.

In many countries, river water has been treated to complement the existing water supply system through bank filtration, such as in Germany, Finland, France, Switzerland, Hungary, and the Netherlands. RBF has become an efficient, well

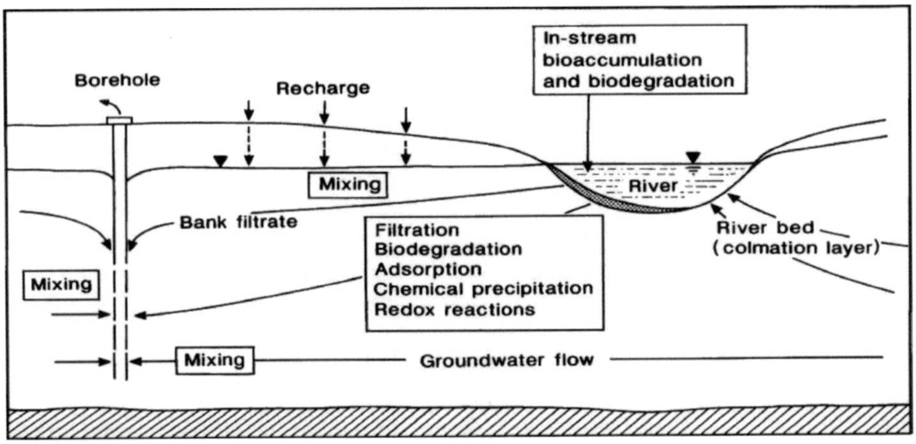

Figure 1.
Diagram of mechanisms in natural filtration by RBF system.

accepted technique for surface water treatment in many European countries. In Switzerland, 80% of the drinking water comes from RBF wells, with 50% in France, 48% in Finland, 40% in Hungary, 16% in Germany, and 7% in the Netherlands. Recently, other countries like Malaysia, India, as well as China and South Korea have started implementing RBF for drinking water supply [22].

Generally, RBF wells constructed in aquifers primarily consist of sand and gravel, with thin layer of granular aquifers (clay or silty sands). Removal of pollutants using RBF involves complex biological, hydrological, and geochemical activities through the aquifer layer during the filtration and infiltration of water. These processes consist of physical filtration, dilution, microbial degradation, precipitation, and sorption processes [1, 16]. According to recent studies in Europe and America, the RBF process is able to provide appropriate defence against microbial contaminants, and reduce the possibility of disinfection by-product formation. Additionally, RBF is also an effective method of removing common microbial pathogens such as *E. coli*, *Microcystins*, and *Cryptosporidium* during the infiltration process [22].

1.2.1 E. coli removal via riverbank filtration

RBF is a water treatment technology that involves extracting water from rivers by pumping wells located in the adjacent alluvial aquifer. In the underground passage, a series of physical, chemical, and biological processes take place, improving the quality of the surface water, while substituting or reducing conventional drinking water treatments. A study based on a model-oriented approach by Wang et al. used an example of riverbank wells near the Kuybyshev Reservoir, Russia [40]. The wells were designed in order to minimise the uncertainties in the estimated hydraulic parameters. During water transport towards the RBF wells, the water quality improved significantly, aided by processes like microbial degradation, ion exchange, precipitation, sorption, filtration, dispersion, and groundwater dilution.

Faecal and total coliforms are bacterial indicators that are widely used to monitor microbial water quality in developed and developing regions of the world. Faecal contamination of drinking water supplies is a public-health concern because they could contain pathogens that cause gastroenteritis, meningitis, and other waterborne diseases [37]. Potential sources of faecal contamination include direct discharge from human and animal wastes as well as non-point sources (agricultural and storm water runoffs). Majority of the RBF systems used in European countries and America alike have achieved excellent total coliform removal percentages, ranging from 99.2 to 99.99% (2.1–5 logs).

Removal of *E. coli* can stand between 99.9 and 99.994% (3–4.2 logs). It was also observed that 25–87% of groundwater can be co-extracted from these RBF wells [15]. According to the compiled literature above, the obtained results revealed the efficiency of the RBF system in eliminating bacteria mainly at the water-media interface. However, in a few situations, the problem of exceeding the stipulated bacterial limits persists. Therefore, it is advisable for the RBF system to be considered as a pre-treatment method to be combined with more effective conventional disinfection technologies in order to meet the target.

1.3 Principles of electromagnetic field treatment

Extremely low-frequency electromagnetic fields (ELF-EMFs) are ubiquitously present in various environments in everyday life. Generally the ELF-EMF spectrum is defined by frequencies from 3 to 3000 Hz [41]. These fields are generated via high-tension electrical distribution networks from residential and occupational sources by power lines and electrical devices. Normally, electric and magnetic fields occur together, and both fields weaken with increasing distance from the source. However, both these fields produce different effects on living organisms. In a large part of the world and Europe, 50 and 60 Hz (in the U.S.) sine wave signals resemble the household alternating current electrical power supply.

The principles and behavioural effects of electromagnetic fields (EMFs) have been reported since the 1970s. Recent studies in the field have shown that exposure to electromagnetic and non-ionising radiations can induce numerous biological effects [42]. For example, exposure to ELF-EMFs ranging from 0 to 100 Hz is capable of activating cellular immune responses. Various approaches have fixated on the probability of analogous effects regarding non-ionising radiation. Despite this, there have been an increasing number of studies suggesting that exposure to ELF-EMFs can affect and slow down the growth of *E. coli* bacteria. ELF-EMFs were also found to decrease the rate of growth in *E. coli* and *S. aureus*, inhibit the growth of cancer cells, and increase the rate of regeneration of worms.

1.3.1 Low-frequency electromagnetic fields

Low-frequency electromagnetic fields (LF-EMFs) are widely applied in electrical appliances and different equipment such as television sets, computers, and kitchen appliances. EMFs are classified into seven categories: (1) Extremely low frequency (0–300 Hz) used in biological processes, (2) low frequency (300 Hz–30 kHz), (3) middle frequency (30 kHz–30 MHz) used for amateur radio and remote controls, (4) ultrahigh (30–300 MHz) used in radio and TV, (5) super high (300 MHz–30 GHz) used in satellite communication, (6) extremely high frequency (30–300 GHz) used in radars, and (7) infrared (300 GHz–300 THz) and visible light (429–750 THz) used in light spectrums. The EMFs are characterised by a frequency of 50 or 60 Hz, and thus occupy the extremely low frequency (ELF), non-ionising range of the electromagnetic spectrum (3 Hz to 3 × 103 Hz) [43]. Although ELF-EMFs do not break molecular bonds or heat body tissue, they may interact with the human body through the weakly generated electric currents.

The LF-EMF is an effective technique in water treatment to prevent scale formation, and detach already-formed scale in industrial water systems. Many related studies have been conducted in the past 60 years, and several devices have been developed. Some researchers have found that EMFs can be implemented in soil and agriculture wastewater disinfection, in therapeutic practices, and in food protection technologies. It was also found that the EMFs have potential in controlling and removing bacterial growth on water treatment systems.

General operations of EMFs in water treatment systems involve the physics of interaction between a magnetic field and moving electric charges. The electromagnetic system consists of magnetic fields generated by coils wrapped around a pipe. The small electrical device treats the water with a patented technology by inducing variable electric currents at a continuous frequency to generate the magnetic field. The magnetic field removes the bacteria (*E. coli*) in the water, and reduces their growth. In other words, magnetic fields interact with the bacteria by the forces and electric field vectors generated. The EMF water treatment system for bacterial removal is presented in **Figure 2**.

1.3.2 Low-frequency electromagnetic fields for E. coli removal

Over the past few decades, it has been well established that non-thermal, non-ionising, extremely low frequency (<300 Hz) and low amplitude (0.2–20 mT) EMFs cause a number of different biological effects. The radiation is said to be capable of inducing physiological effects too [44]. Several studies have been performed to verify the direct effects exerted by EMFs on cellular functions. Bacteria were also tested in some studies using the ELF-EMF wave. The biological effects of EMFs are quite heterogeneous, depending on the types of cell studied, intensity, and types of field used. For more than three decades, various forms of electrical stimulation, including capacity coupling, direct current, and combined magnetic fields have been used as a therapeutic remedy. The current, and one of the most useful methods to investigate antibacterial effects due to magnetic fields is to use low-frequency EMFs, especially frequencies ranging from 50 to 60 Hz. On the contrary, even with some publications claiming the bio-effects of EMFs, there are plenty of studies showing no significant effects on living organisms. Podda et al. reported no change in oxidative DNA damage after 50 Hz EMF exposure was applied [45].

The implementation of magnetic fields in water treatment systems is a non-chemical method that covers a wide range of technologies. A recent study reported that EMFs had a positive effect on the efficiency in the number of bacteria removed in biological wastewater treatment [46]. EMFs also intensify the stationary-phase-specific transcription activity of the *E. coli* bacteria [47]. Additionally, Piydasa et al. observed growth responses of healthy *E. coli* cells exposed to EMF energy for 7 hours through the water treatment chamber [48]. They indicated a statistically

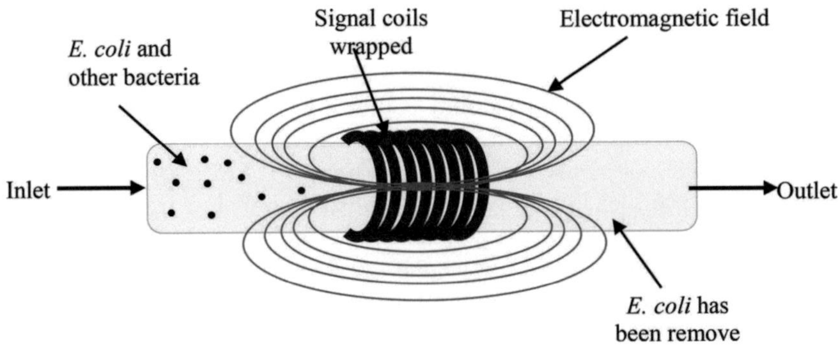

Figure 2.
Electromagnetic system for water treatment consists of magnetic field generated by coils wrapped around the pipe.

significant inhibition in the growth of *E. coli* that was attributed to the effect of different EMF waveforms and applied energies between two pulsed-electromagnetic field devices, Device D and Device G. Therefore, the stimulation or inhibition of the growth of *E. coli* and other bacteria depends on the frequency, field strength, and type of microorganism.

The data on LF-EMFs on *E. coli* effects for different environmental conditions, such as in water suspension, is important for future applications. As a major constituency, water is a medium for biological systems as well. The LF-EMF effects on H_2O molecules can be the mechanism in the creation of conditions for biological responses [49]. Different effects on the growth and viability of *E. coli* after bacterial suspension exposure to LF-EMFs have been reported at several frequencies that are resonant for H_2O molecules. According to Belyaev, LF-EMFs can cause direct physical damage, ionisation, or heating of *E. coli*, resulting in morphological changes after exposure [50].

1.3.3 Design and construction of LF-EMF exposure column pipe

The initial LF-EMF column structure was designed using the ANSYS Maxwell software program. A geometrical design of the model was created and drawn with the selected parameters. A fully-automatic meshing procedure was applied before the simulation was begun, after the structure had been modelled. In the ANSYS Maxwell 3D, the solutions were based on meshes by using thousands of LF-EMF coiled column elements. Accurate solutions based on coarse meshes using relatively few elements were obtained. To assist the meshing, the coil workpiece was created with multi-layers depending on the skin depth. The aim of this simulation was to generate a uniform magnetic field inside the column pipe, and assign the number of coil turns for the LF-EMF coiled column. **Figure 3** illustrates the cross-section structure of the LF-EMF coiled column, as well as the dimensions for one of the five column test models in this study.

There were five models of LF-EMF coiled columns built using ANSYS Maxwell software with different designs in this study. All the LF-EMF coiled columns were designed using gauge insulated copper wire with the conductivity of wound wires in a vacuum environment. The diameter range of the copper wires used was 1.5 mm. This range coil diameters were used to determine the most effective diameter of coil in generating a magnetic field at different column diameters (D_{column}), and number of coil turns (N). Theoretically, increasing the number of coil turns, with the same coil and the same current flowing, would increase the magnetic field strength. This relationship is defined as magnetomotive force (MMF), which refers to the flowing

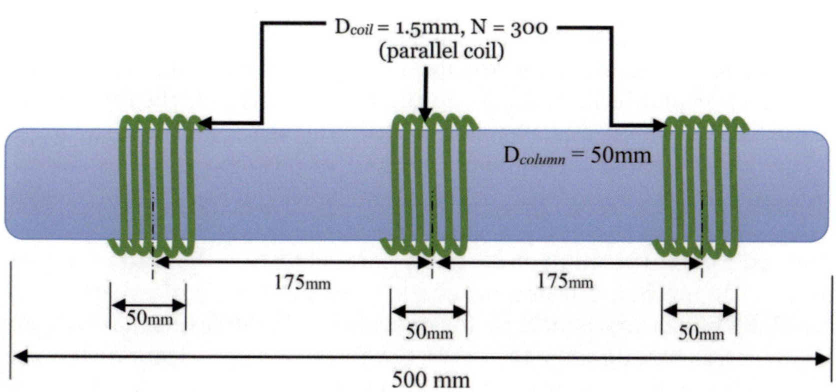

Figure 3.
LF-EMF column coil structure with workpiece dimensions.

of current through a coil of N turns. Therefore, an electromagnetic field strength can be determined by the *ampere-turns* of the coil; the more turns of wire in the coil, the greater the strength of the magnetic field.

The simulations of the coiled columns began with determining the N using 2 A excitation current to generate a magnetic field in the range of 2–10 mT. The magnetic field generated by the electromagnet was proportional to both N and *I* in the coil winding. Coil winding revolves around the geometry of the wound coil. In the opposite position of the wound coil is a calculated coil structure within the winding space. In this study, the N value and length of coil were designed depending on the length of solenoid, type of coil, and column diameter. The parameters of required number for coil turns (N), and diameter of coil were calculated using Maxwell-Ampere's law. The winding of the coil was manually wrapped parallel to the column at different diameters of coil and column. The coils were positioned in parallel to the column to generate a homogeneous horizontal LF-EMF coiled column with the water sample centred at the horizontal axis of the coil pairs.

The strength and intensity of the LF-EMF coiled columns depends on the N of coils, diameter of coils, and also the type of column material used as the core. In this study, five transparent polypropylene cylindrical columns with inner diameters of 50 mm, which were 500 mm in length were used. These cylindrical columns were of non-magnetic material that can be regarded as free space as they have a very low value of permeability. This material had no effect on concentrating the magnetic flux, and the magnetic field created by the current in the coils. The performance parameters of the coil used was dependent the geometry and coil dimension ratios. A smaller diameter of coil required much more number of coil turns, while bigger coil diameters used resulted in larger magnetic fields.

2. Material and methods

LF-EMF was produced with the coiled column and the sinusoidal 50 Hz magnetic field was generated by means of a solenoid, obtained with overlapped winding of copper wire with 500 mm of length and 50 mm diameter of column A and 80 mm diameter of column B. The power of 220 V was connected to solenoid coils waveform generator. The current flowing in the exposure devices was monitored by AC current in parallels of coils. In each coil, the number of turns was between 300 and 600 of 1.5 mm copper wire and the total resistance of the system was 1 Ω. The resulting of the whole inductance coil system was in the range of 2–10 mT. The density of magnetic flux was monitored using Tesla meter (BST600 Gauss meter/ Tesla Meter) and field intensity was varied by ±0.05 mT.

According to previous studies, the method of IDEXX Colilert® 18 has been verified as an acceptable alternative to other test methods for the recovery of *E. coli* from source water, wastewater, and drinking water [51, 52]. The original method using mTEC agar is a standard method, and is recognised by the U.S.EPA as a useful method for monitoring recreational water quality [52, 53]. In this study, the Colilert® 18 method was compared to the original mTEC membrane filter method for enumeration of *E. coli* by parallel testing the water samples. The Colilert® 18 method was determined to be an acceptable alternative to the traditional mTEC standard method for monitoring *E. coli* levels in river water recreational areas based on the results of the study. No significant difference was found between the results of these two methods. Therefore, the Colilert® 18 method was suggested as an efficient and accurate means for testing water samples in this study.

2.1 Sample characterisation

The study site is located on Sungai Kerian at Lubok Buntar Kedah, Malaysia (**Figure 4**). The sample was collected during dry and wet seasons from Sungai Kerian and tube well. One hundred samples were collected; 50 sample for column test. Tube well water samples were collected in sterile amber glass bottles of 125 ml without headspace in order to prevent the formation of air bubbles and were airtight sealed. All laboratory analytical analysis was according to Standard Method [37]. Ten litres of river raw water sample was collected in polyethylene bottles. These samples were preserved in accordance with Water and Wastewater Standards and then stored at a temperature of less than 4°C. Laboratory apparatus used in this study were prewashed with 5% nitric acid (HNO_3) and rinsed with deionised water prior to testing.

2.2 Experimental setup

In this study, 13 runs of test with different magnetic field exposures were conducted. The horizontal transparent polypropylene cylindrical columns test with inner diameters of 50 mm were used. The coiled column test was built on the horizontal axis, and the magnetic field was generated by means of a solenoid, obtained with overlapping windings of copper wires, at 500 mm in length. The coils wrapped around the column transformed the magnetic fields controlled by the magnetic field power generating system. The maximum of effective current was 2 A at 50 Hz frequency. The magnetic generator consisted of a pair of solenoid coils, a current amplifier, and a waveform generator controller. The samples were exposed and placed in a horizontal column where uniformity of the magnetic field was optimal. Then, the LF-EMF coiled column experiment was started by pumping water samples in a horizontal direction into the column in order to ensure complete wetting of the particles [54]. A constant discharge rate was maintained using the peristaltic pump model Masterflex L/S HV 07522–20 at 50 and 100 mL/min to allow sufficient contact time between the magnetic field, and for the removal of *E. coli*.

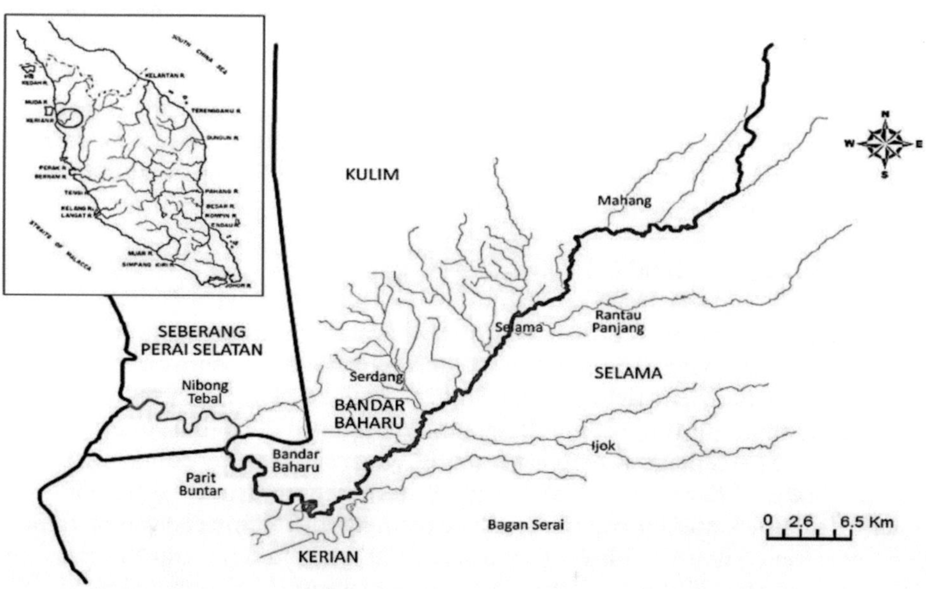

Figure 4.
The location of Kerian River in map of peninsular Malaysia.

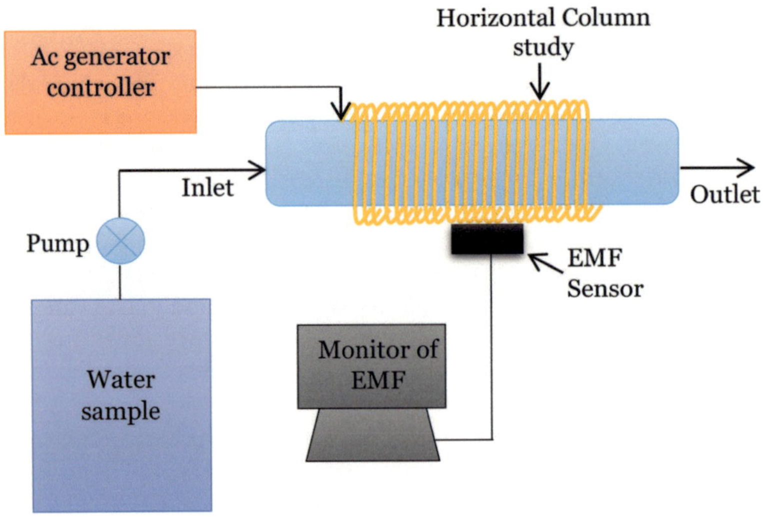

Figure 5.
Set-up for LF-EMF column study on E. coli removal by using magnetic fields exposure with alternating current.

The water sample was continuously fed into the column for a total experiment time of 6 hours. A total of 18 effluents were collected every hour directly into 100 mL sterilised vessels. The experimental setup is shown in **Figure 5**.

3. Result and discussion

Data collected from the experiments were analysed using Microsoft Excel and Origin-Pro 9.1 software. All the data and results are presented in the form of tables and figures. The findings include the simulation results and validation of the LF-EMF exposure, water sample characteristics, *E. coli* concentration in water quality, removal mechanisms, and potential of the LF-EMF column to remove *E. coli*. The results from the comparative analysis of *E. coli* removal before and after LF-EMF exposure with selected column diameters, and magnetic flux densities of 2 to 10 mT are presented in this chapter. In the present study, the optimisation of *E. coli* removal contact time for LF-EMF exposure was performed by using response surface methodology (RSM) to evaluate the optimal magnetic flux design, geometry, and parameters during the designing process.

3.1 Effects of magnetic field (β)

The column experiments consisted of LF-EMF coiled columns and river water, which was varied to analyse the effects of the magnetic field exposure on the removal of *E. coli*. Varying magnetic field intensities of 2, 4, 6, 8 and 10 mT reacted with the water samples at 50 and 100 mL/min flowrates. The data on the initial concentration of *E. coli* and the percentage of removal from the water samples were obtained.

Removal rates of *E. coli* in the water samples were measured at five different magnetic field intensities for column test. The diameter of column test was 50 mm, which underwent 6 hours of LF-EMF exposure at 50 mL/min (Q1), and 100 mL/min (Q2) water flowrates. **Figure 6** illustrates the percentage of *E. coli* removal after LF-EMF exposure. The maximum removal efficiency of *E. coli* in column test was 100% at 6, 8, and 10 mT of magnetic field exposure for Q_1. However, percentage of

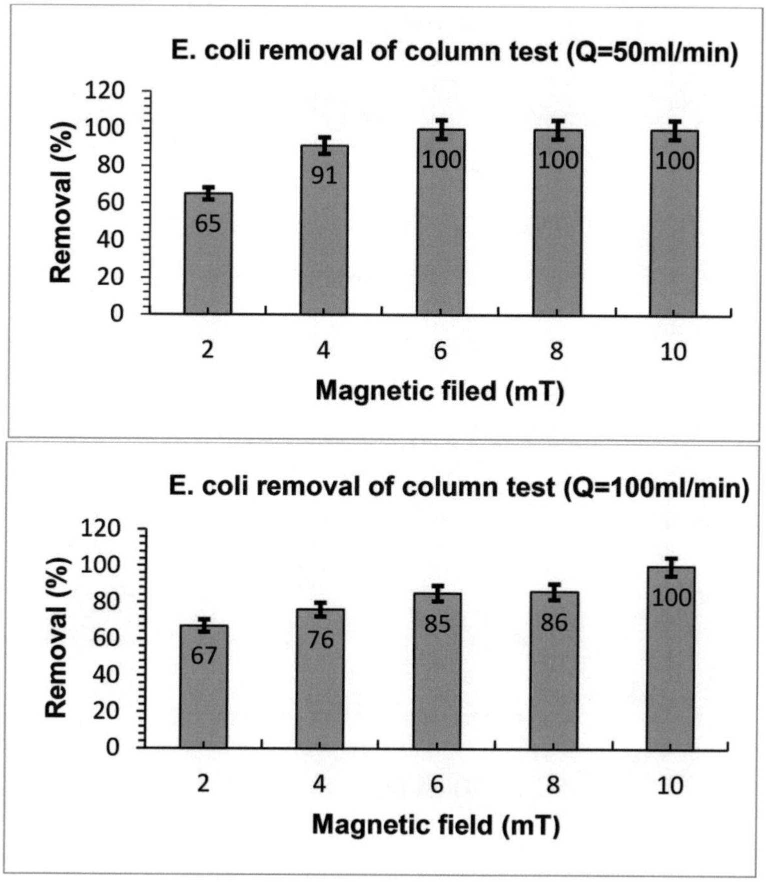

Figure 6.
Percentage of E. coli removal of column test (⌀ = 50 mm) for Q_1 and Q_2.

E. coli removal for Q_2 was 67% at 2 mT, and 100% of removal at 10 mT of magnetic field exposure. These results indicated that the *E. coli* in the sample of water that was exposed to the LF-EMF was statistically significantly decreased. The magnetic intensity of the LF-EMF changed the characteristic responses for *E. coli* bacteria [24].

From **Figure 6**, the results obtained demonstrated that column test achieved 100% *E. coli* removal, which increased with the increase of magnetic field exposure at Q_1 and Q_2. These results indicated that the magnetic field intensity was affected by the surface area of the coiled column. The 50 mm diameter of column test was effective for *E. coli* removal by LF-EMF exposure. This situation is possibly due to the influence of magnetic flux at various flow velocities, where the magnetic field killed the *E. coli* during the exposure at the surface area of the column. Thus, these results indicated that the removal of *E. coli* in column test was 100%, and also showed that the increase of *E. coli* removal is dependent on the increase in intensity of magnetic field along with other parameters of the column.

3.2 Effect of contact time in column test

In order to investigate the effect of LF-EMF exposure for varying durations of 2–6 hours on the removal of *E. coli*, the treatment conditions of exposure intensity were varied from 2 to 10 mT LF-EMF in column test. **Figure 7** illustrates the experimental results of the LF-EMF exposure hourly. The results demonstrated that the removal of *E. coli* approximately increased with the longer of exposure time. The

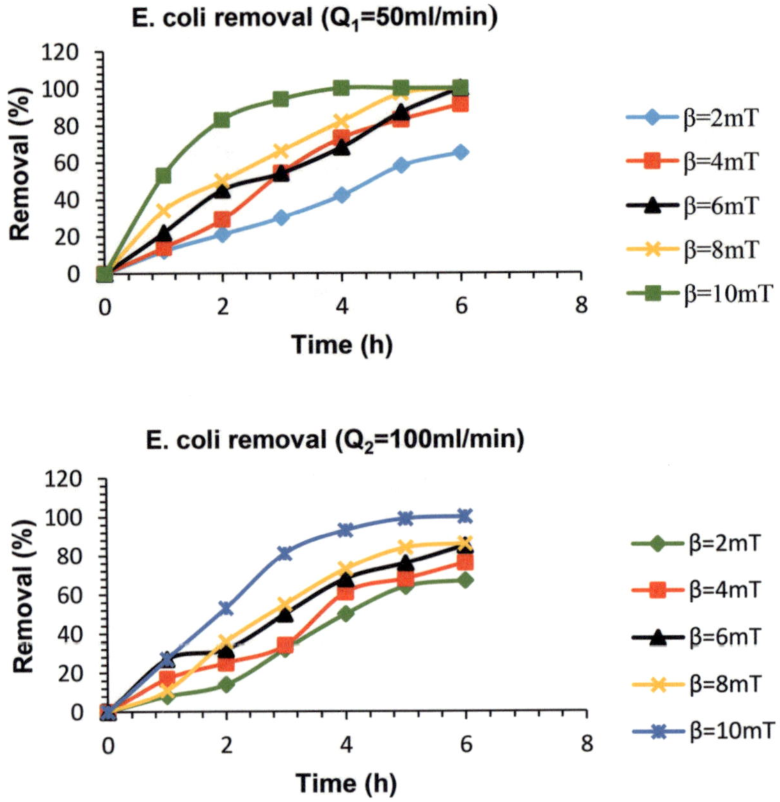

Figure 7.
Effect of time on removal of E. coli (Ø = 50 mm).

percentages of removal in column test achieved 100% at 4 hours of treatment at 10 mT of magnetic field intensity at Q_1 and Q_2. **Figure 7** is seen below.

The optimal time of magnetic field exposure on column test for *E. coli* removal is shown in **Figure 7**, which was at 4–6 hours exposure for both Q_1 and Q_2. The removal was 100% from 4 to 6 hours of 10 mT of LF-EMF exposure. These results indicated that the velocity (*V*) of water samples through column test was effective. Therefore, the removal rates of *E. coli* were constant after 4 hours of contact time with the magnetic field for Q_1 and Q_2. The magnetic field decreased the concentration of *E. coli*, and thus suggests that it affects the behaviour of *E. coli*. Strašák et al. reported that the inhibitory effects on *E. coli* concentration were increased with length of exposure [55]. Also Gaafat et al. applied extremely LF-EMF for 6 and 16 hours on the *E. coli*, and found that an exposure period of 6 hours decreased the concentration of *E. coli*, but after a 16-hour exposure period, they became more resistant towards it.

The effect of the exposure to LF-EMF at 2–10 mT on *E. coli* removal was determined by experimental results performed on the entire data for column test with water flowrates of Q_1 and Q_2. This effect generally depended on the magnetic field intensity and time of exposure [56], whereby the significant effect on *E. coli* removal was increased according to exposure time. From the experimental study, the R^2 results gave practical importance on exposure time for the removal of *E. coli*. From the results, it was found that the removal of *E. coli* and exposure time on column test were significantly correlated with high R^2 values for 2–8 mT of LF-EMF. However, at 10 mT of magnetic field intensity, the non-linear regression with polynomial R^2 at second order for Q_1 was 0.9721, while R^2 for Q_2 was 0.9958. These results showed that the *E. coli* removal at 10 mT exposure was 100% after 4 hours of exposure.

4. Conclusion

The ability of LF-EMFs to remove or decrease the concentration of *E. coli* in the river water samples was successfully demonstrated in this study. Based on the results, LF-EMFs were able to kill a part of the *E. coli*, and decreased the concentration by magnetic field exposure. From this study, the LF-EMF proved its capability to remove and control the growth of *E. coli* with magnetic field exposure. The effect of the magnetic field in the removal of *E. coli* by using an LF-EMF column model was validated with experimental results. Initially, simulations were carried out to study and design the magnetic field generating system, and compare the results with previous experiments. This result indicates that the application of the LF-EMF coiled column with a magnetic field at 6 mT was able to remove 100% of *E. coli*. The surface area and volume of the column induced changes in the percentage of *E. coli* removal. Therefore, it was found that the LF-EMF column test was effective column size to remove *E. coli* from the water. Data from the monitoring study for the RBF tube well showed low concentrations of *E. coli* during the wet and dry seasons. Thus, the result showed the suitability of LF-EMF column application in RBF as an alternative technique to control *E. coli* growth. Other than that, LF-EMF technology, as a non-chemical and non-ionising technique, was proposed in this study to increase the quality of water in RBF especially during the wet season. Some technical and fundamental principles related to the application of this technique provided valuable information about its capabilities in drinking water treatment applications, and at the same time, provided other opportunities for further research.

Acknowledgements

The authors thank the Ministry Education Malaysia for providing LRGS Grant on Water Security entitled Protection of Drinking Water: Source Abstraction and Treatment (203/PKT/6726001). This project was also partly supported by USM Research University Individual (RUI) Grant (1001/PAWAM/814287).

Author details

Rossitah Selamat[1], Ismail Abustan[1*], Mohd Rizal Arshad[2] and Nurul Hana Mokhtar Kamal[1]

1 Faculty of Civil Engineering, Universiti Sains Malaysia, Pulau Pinang, Malaysia

2 Faculty of Electrical Engineering, Universiti Sains Malaysia, Pulau Pinang, Malaysia

*Address all correspondence to: ceismail@usm.my

References

[1] Umar DA. An overview assessment of the effectiveness and global popularity of some methods used in measuring riverbank filtration. Journl of Hydrology. 2017;**550**:497-515

[2] Guidelines for Drinking-water Quality, 4th edition. Geneva: World Health Oragnisation (WHO); 2011

[3] Abia ALK, Ubomba-Jaswa E, Momba MNB. Impact of seasonal variation on *Escherichia coli* concentrations in the riverbed sediments in the Apies River, South Africa. Science of the Total Environment. 2015;**537**:426-469

[4] Azlan A, Khoo HE, Idris MA, Ismail A, Razman MR. Evaluation of minerals content of drinking water in Malaysia. The Scientific World Journal. 2012;**2012**:1-10

[5] Adlan MN, Ghazali MF. Removal of *E. Coli* and turbidity using riverbank filtration technique (RBF) for riverside alluvial soil in Malaysia. Journal of the Institute of Engineers, Malaysia. 2016;**77**(1):30-35

[6] Coffey R, Benham B, Krometis LA, Wolfe ML, Cummins E. Assessing the effects of climate change on waterborne microorganisms: Implications for EU and U.S. water policy. Human and Ecological Risk Assessment: An International Journal. 2014;**20**(3):724-742

[7] World Health Organisation (WHO). The Treatment of Diarrhoea. 2005

[8] Choi CJ, Berges JA, Young EB. Rapid effects of diverse toxic water pollutants on chlorophyll a fluorescence: Variable responses among freshwater microalgae. Water Research. 2012;**46**(8):2615-2626

[9] Garzio-Hadzick A, Shelton DR, Hill RL, Pachepsky YA, Guber AK, Rowland R. Survival of manure-borne *E. coli* in streambed sediment: Effects of temperature and sediment properties. Water Research. 2010;**44**(9):2753-2762

[10] Amin MT, Alazba AA, Manzoor U. A review on removal of pollutants from water/wastewater using different types of nanomaterials. Advances in Materials Science and Engineering. 2014;**2014**:825910

[11] Buzek F, Kadlecova R, Jackova I, Lnenickova Z. Nitrate transport in the unsaturated zone: A case study of the riverbank filtration system Karany, Czech Republic. Hydrological Processes. 2011;**26**:640-651

[12] Sandhu C, Grischek T, Kumar P, Ray C. Potential for riverbank filtration in India. Clean Technologies and Environmental Policy. 2011;**13**(2):295-316

[13] Schubert J. Significant of hydrologic aspects on RBF performance. Riverbank Filtration Hydrology. 2006;**38**:1-20

[14] Kuehn W, Mueller U. Riverbank filtration: An overview. Journal-American Water Works Association. 2000;**92**(12)

[15] Levantesi C et al. Quantification of pathogenic microorganisms and microbial indicators in three wastewater reclamation and managed aquifer recharge facilities in Europe. Science of the Total Environment. 2010;**408**(21):4923-4930

[16] Wang P, Pozdniakov SP, Shestakov VM. Optimum experimental design of a monitoring network for parameter identification at riverbank well fields. Journal of Hydrology. 2015;**523**:531-541

[17] Bajpai I, Saha N, Basu B. Moderate intensity static magnetic field has bactericidal effect on *E. coli* and *S. epidermidis* on sintered hydroxyapatite. The Journal of Biomedical Materials

Research Part B: Applied Biomaterials. 2012;**100**(5):1206-1217

[18] Modified FU. (*E. coli*) in water by membrane *Escherichia coli* agar (Modified mTEC). U.S. Environmental Protection Agency. 2002;**821**:2-23

[19] Bauer R, Dizer H, Graeber I, Rosenwinkel K, Lo JM. Removal of bacterial fecal indicators, coliphages and enteric adenoviruses from waters with high fecal pollution by slow sand filtration. Water Research. 2011;**45**(2):439-452

[20] Zietzschmann F, Aschermann G, Jekel M. Comparing and modeling organic micro-pollutant adsorption onto powdered activated carbon in different drinking waters and WWTP ef fl uents. Water Research. 2016;**102**:190-201

[21] Tijani JO, Fatoba OO, Madzivire G, Petrik LF. A review of combined advanced oxidation technologies for the removal of organic pollutants from water. Water, Air, & Soil Pollution. 2014;**225**(9):2102(1-46)

[22] Hu B, Teng Y, Zhai Y, Zuo R, Li J, Chen H. Riverbank filtration in China: A review and perspective. Journal of Hydrology. 2016;**541**:914-927

[23] Oncul S, Cuce EM, Aksu B, Inhan Garip A. Effect of extremely low frequency electromagnetic fields on bacterial membrane. International Journal of Radiation Biology. 2016

[24] Tessaro LWE, Murugan NJ, Persinger MA. Bacterial growth rates are influenced by cellular characteristics of individual species when immersed in electromagnetic fields. Microbiological Research. 2015;**172**:26-33

[25] Azrina A, He K, Ma I, Amin I. Major inorganic elements in tap water samples in peninsular Malaysia. Malaysian Journal of Nutrition. 2011;**17**(2):271-276

[26] Naubi I, Zardari NH, Shirazi SM, Ibrahim NFB, Baloo L. Effectiveness of water quality index for monitoring Malaysian river water quality. Polish Journal of Environmental Studies. 2016;**25**(1):231-239

[27] Hafiza N, Razak A, Praveena SM, Hashim Z, Zaharin A. Drinking water studies: A review on heavy metal, application of biomarker and health risk assessment (a special focus in Malaysia). The Journal of Epidemiology and Global Health. 2015;**5**(4):297-310

[28] Ong C, Ibrahim S, Sen Gupta B. A survey of tap water quality in Kuala Lumpur. A survey of tap water quality in Kuala Lumpur. Urban Water Journal. 2007;**4**(1):29-41

[29] Yuk Feng Huang TSL, Ang SY, Lee KM. Quality of water resources in Malaysia. 2015

[30] Mwabi JK, Mamba BB, Momba MNB. Removal of waterborne bacteria from surface water and groundwater by cost-effective household water treatment systems (HWTS): A sustainable solution for improving water quality in rural communities of Africa. Water SA. 2013;**39**(4):445-456

[31] Kostyla C, Bain R, Cronk R, Bartram J. Seasonal variation of fecal contamination in drinking water sources in developing countries: A systematic review. Science of the Total Environment. 2015

[32] Ishii S, Ksoll WB, Hicks RE, Sadowsky MJ. Presence and growth of naturalized *Escherichia coli* in temperate soils from Lake Superior watersheds. Applied and Environmental Microbiology. 2006;**72**(1):612-621

[33] José Figueras M, Borrego JJ. New perspectives in monitoring drinking water microbial quality. International Journal of Environmental Research and Public Health. 2010;7(12):4179-4202

[34] Revetta RP, Gomez-Alvarez V, Gerke TL, Santo Domingo JW, Ashbolt

NJ. Changes in bacterial composition of biofilm in a metropolitan drinking water distribution system. Journal of Applied Microbiology. 2016;**121**(1):294-305

[35] A. Public and H. Association. APHA method 4500-CL: Standard methods for the examination of water and wastewater. 2005

[36] Shoaib M et al. Prevalence of pathogenic microorganisms in drinking water of Rawalpindi and Islamabad. World Journal of Fish and Marine Sciences. 2016;**8**(1):14-21

[37] World Health Organisation (WHO). Guidelines for Drinking-Water Quality. Vol. 1. 2006

[38] Saxena T, Kaushik P, Krishna Mohan M. Prevalence of *E. Coli* O157: H7 in water sources: An overview on associated diseases, outbreaks and detection methods. Diagnostic Microbiology and Infectious Disease. 2015;**82**(3):249-264

[39] Odonkor ST, Ampofo JK. *Escherichia coli* as an indicator of bacteriological quality of water: An overview. Microbiology Research (Pavia). 2013;**4**(1):5-11

[40] Wang Z, Xiao G, Zhou N, Qi W, Han L, Ruan Y. Comparison of two methods for detection of fecal indicator bacteria used in water quality monitoring of the three gorges reservoir. Journal of Environmental Sciences. 2015;**38**:42-51

[41] Angelo CD. Experimental model for ELF-EMF exposure: Concern for human health. Saudi Journal of Biological Sciences. 2015;**22**(1):75-84

[42] Nasri K, Daghfous D, Landoulsi A. Effects of microwave (2.45 GHz) irradiation on some biological characters of *Salmonella typhimurium* Effets des micro-ondes de fre'quence 2,45 GHz Sur lescaractères biologiques de *Salmonella typhimurium*. Comptes Rendus Biologies. 2013;**336**:194-202

[43] Egervärn M, Englund S, Ljunge M, Wiberg C, Finn M. Unexpected common occurrence of transferable extended spectrum cephalosporinase-producing *Escherichia coli* in Swedish surface waters used for drinking water supply. Science of the Total Environment. 2017

[44] Martirosyan V, Baghdasaryan N, Ayrapetyan S. Bidirectional frequency-dependent effect of extremely low-frequency electromagnetic field on *E. coli* K-12. Electromagnetic biology and medicine. 2013;**32**(3):291-300

[45] Podda MV et al. Extremely low-frequency electromagnetic fields enhance the survival of newborn neurons in the mouse hippocampus. The European Journal of Neuroscience. 2014;**39**(6):893-903

[46] Inhan-Garip A, Aksu B, Akan Z, Akakin D, Ozaydin AN, San T. Effect of extremely low frequency electromagnetic fields on growth rate and morphology of bacteria. International Journal of Radiation Biology. 2011;**87**(12):1155-1161

[47] Zhang J et al. A new integrated potable reuse process for a small remote community in Antarctica. Process Safety and Environment Protection. 2016;**104**:196-208

[48] Piyadasa C et al. The effect of electromagnetic fields, from two commercially available water treatment devices, on bacterial culturability. Water Science and Technology. 2016;**73**(6):1371-1377

[49] Torgomyan H, Trchounian A. Low-intensity electromagnetic irradiation of 70.6 and 73 GHz frequencies enhances the effects of disulfide bonds reducer on *Escherichia coli* growth and affects the bacterial surface oxidation-reduction

state. Biochemical and Biophysical Research Communications. 2011;**414**(1):265-269

[50] Belyaev I. Withdrawn: Toxicity and SOS response to ELF magnetic field and nalidixic acid in *E. coli* cells. Mutation Research–Genetic Toxicology and Environmental Mutagenesis. 2011;**722**(1):84-88

[51] Edberg SC, Rice EW, Karlin RJ, Allen MJ. *Escherichia coli*: The best biological drinking water indicator for public health protection. Journal of Applied Microbiology. 2000

[52] Kinzelman JL, Singh A, Ng C, Pond KR, Bagley RC, Gradus S. Use of IDEXX colilert-18® and quanti-tray/2000 as a rapid and simple enumeration method for the implementation of recreational water monitoring and notification programs. Lake and Reservoir Management. 2005;**21**(1):73-77

[53] Budnick GE, Howard RT, Mayo D, editors. Comparison of Colilert-18 to the mTEC agar method for the enumeration of *Escherichia coli* in recreational waters. Hartford , Connecticut: Mayo Division of Laboratory Services, Connecticut Department of Public Health; 2001

[54] Dastanaie AJ, Bidhendi GRN, Nasrabadi T, Habibi R, Hoveidi H. Use of horizontal flow roughing filtration in drinking water treatment. International journal of Environmental Science and Technology. 2007;**4**(3):379-382

[55] Strašák L, Vetterl V, Šmarda J. Effects of low-frequency magnetic fields on bacteria *Escherichia coli*. Bioelectrochemistry. 2002;**55**(1-2):161-164

[56] Nawrotek P, Fijałkowski K, Struk M, Kordas M, Rakoczy R. Effects of 50 Hz rotating magnetic field on the viability of *Escherichia coli* and *Staphylococcus aureus*. Electromagnetic Biology and Medicine. 2014;**33**(1):29-34

Chapter 7

Photocatalytic Treatment of Pesticides Using TiO_2 Doped with Rare Earth

Juan C. Arévalo Pérez, José Gilberto Torres Torres, Durvel de la Cruz Romero, Hermicenda Perez-Vidal, Maria Antonia Lunagomez Rocha, Ignacio Cuauhtémoc López, Adrian Cervantes Uribe and Zenaida Guerra Que

Abstract

Rare earth doping ions can improve the spectral response of this semiconductor to the visible region. This work evaluated the dopant effect of rare earth ions such as La, Ce, Nd, Pr, Sm, Eu, and Gd in titania for the solar photodegradation of Diuron and methyl parathion. The increase in the content up to 0.5% of dopants decreases photoactivity due to the formation of photo-generated electron-hole pair recombination centers. The catalysts calcined at 500°C presented only the anatase crystalline phase and the samples doped with La and Ce at 0.1 and 0.3% were the most active in diuron solar degradation; however, when the temperature of the thermal treatment increased to 800°C, mixtures of crystalline phases were presented. The catalyst with the highest anatase content showed the best performance. The materials calcined at 500°C with better performance in diuron solar degradation were selected to to treat methyl parathion using solar light. Finally, under these conditions, an affinity was found for the dopant ions in titania and in the functional groups of the contaminating molecules (phenylurea and thiophosphate). Solar photodegradation of diuron was more effective with La and Ce, while for methyl parathion, it was Eu at 0.3%.

Keywords: rare earth ions, doped TiO_2, diuron, methyl parathion, sunlight

1. Introduction

Currently, the use of pesticides has increased in order to eliminate the pests that limit and reduce agricultural production in all countries of the world. Consequently, this has caused these substances to run off into natural aquatic bodies contaminating this medium. Diuron and methyl parathion have manifested this problem; hence, it is possible to find concentrations of these pesticides in aquatic bodies close to where they are applied [1]. Although the solubility in water is low, they can dissolve due to the surrounding environment. The presence of diuron and methyl parathion in water is difficult due to the persistence and stability they present [2].

In recent years, the elimination of these compounds in water has been reported by methods, physical [3], biological [4], and chemical [5]. Of the latter, advanced oxidation processes, such as heterogeneous photocatalysis, have been shown to be very efficient in the chemical transformation of pollutants up to their mineralization to CO_2 and other harmless compounds [6].

This process starts when a semiconductor is excited, with light that has a wavelength greater than or equal to the band energy of the semiconductor, to generate electron-hole pairs, which combine with the water and oxygen of the medium to form radicals that oxidize and mineralize the polluting organic matter [7]. TiO_2 is the ideal semiconductor used for this process; unfortunately, its spectral response is carried out at wavelengths corresponding to the UV region, which limits its use with natural sunlight because its spectrum only has a 5% UV light. Therefore, the investigations related to this semiconductor are made to improve its spectral response in the visible region, which has been achieved by doping the titania with different elements such as non-metals [8], transition metals [9], noble metals [10], and rare earth [11]. In this chapter, we analyzed the photocatalytic behavior under the natural sunlight of TiO_2 doped with La, Ce, Nd, Pr, Sm, Eu, and Gd at 0.1, 0.3, and 0.5% by weight thermally stabilized at 500 and 800°C for the degradation of diuron and methyl parathion.

1.1 Rare earth elements in photocatalysis

Rare earth ions have been used for doping TiO_2 aiming to modify the spectral response of the semiconductor to the visible light region to enhance its photocatalytic properties. Specifically, these ions can displace the phase transformation of anatase to rutile due to high temperature. Furthermore, have the capacity to form complexes with various base Lewis such as amines, aldehydes, carboxylic acids, alcohols, thiols etc., by the presence of electrons coming f-orbitals that interact with these functional groups, consequently allows in improving the absorptivity of organic pollutants in the aqueous medium and to elevate the photocatalytic activity [12]. These trivalent ions possess energy levels with a form of stair that as a dopant in a semiconductor can emit UV or visible light, through sequential absorptions from many near-infrared photons. The transformation of light from near-infrared and visible spectra toward UV range can be used to excite band gap of the titania [13].

On the other hand, luminescent properties of rare earth ions are originated by the electronic transition in the f-orbitals, which are partially full. These are sterically shielded from surrounding microenvironment by filled 5 s and 5p orbitals, generating narrow bands with specific emission energy for each rare earth ion. This process provides properties unique to rare earth ions in photocatalytic applications [14].

At the end of the 1990s, the first investigations involving rare earth ions as dopants in TiO_2 for the photocatalytic oxidation were started [15]. Lin and Yu use a commercial photocatalyst (TiO_2-P25) as a semiconductor for the acetone oxidation, doping this material with La. Then many reports appeared describing the doping of TiO_2 with rare earth ions applying methods of preparation such as solvothermal, microemulsion, impregnation, electrospinning, magnetron sputtering, and sol-gel [16]. The latter has been the most used due to its easy process and its low cost.

In previous studies, it has been found that the insertion of rare earth ions such as lanthanum in titania, cannot replace the position of Ti, due to the large size of the lanthanum ion with respect to Ti [17]. Typically, the rare earth ions on the surface of the TiO_2 are adsorbed in the form of oxides; only the titanium surface can be replaced by rare earth ions in the network of adsorbed lanthanide oxides, forming the Ti-O-L bond [18]. However, the substitution of a trivalent rare earth ion by a tetravalent titanium ion creates an imbalance, favoring centers with positive charges, which could adsorb anions such as OH ions, to compensate the charge balance [19]. Therefore, the photo-generated holes can be consumed immediately after the load carriers are transferred to the surface, which increases the efficiency in the separation of charges. The photocatalytic benefits of the anatase and rutile phase in titania are widely known; the addition of rare earth ions in materials with these crystalline phases shows a growth in the crystal size, due to the presence of the Ti-O bonds, in the interface between TiO_2 and rare earth oxide formed [20]. In the interaction with anatase, the presence of these mentioned bonds inhibits the thermal transformation at the critical temperature of change, manifesting mixtures of crystalline phases at temperatures above 700°C in titania.

The rare earth oxides modified with titania show a growth in the intensity of light absorbed compared to pure TiO_2. According to Yan et al., incident photons can be scattered and lost by reflection on a smooth surface, while on a rough surface, formed by the presence of rare earth oxides, allows a large number of scattered photons penetrate the interior of the particle to activate the separation of charges [21].

In inorganic semiconductors such as TiO_2, light absorption is mainly attributed to the transition from the valence band to the conduction band, which is commonly referred to as band transitions. However, it is believed that, in the presence of lanthanide oxides, the increase in the intensity of light absorption is due to the transition of the electrons belonging to layer 4f of the lanthanides, known as the transition $f \rightarrow f$. The corresponding energy can be transferred to the titania to separate the charges [22].

1.2 Pesticides treated by photocatalysis

1.2.1 Diuron

Diuron (3-(3, 4-dichlorophenyl)-1, 1-dimethylurea) is a white, crystalline and odorless powder, which has low solubility in water (36.4 mg/L). Herbicide is employed for weed control in non-crop areas and to control weeds in a range of tree crops. Its mechanism of action mainly acts inhibiting photosynthesis by blocking electron transport at photosystem II [23]. When it is applied to soils, it is leached from 3 to 5 cm and strongly adsorbs persisting up to 330 days. In aqueous medium, it is partially absorbed due to its solubility, by the action of solar photolysis and OH radicals present in the environment, almost completely degraded, but this process is too slow and depends on environmental conditions [24]. For this reason, advanced oxidation processes such as heterogeneous photocatalysis with TiO_2 have been used to eliminate this herbicide as a water pollutant.

When TiO_2 is used as a colloidal particle in an aqueous solution of diuron, only one transformation is observed in the aliphatic chain, where the OH radicals attack the benzene ring causing its opening to aliphatic chains. In the presence of acetonitrile, the reaction mechanism indicates a reductive discoloration of the benzene ring, without it an oxidative demethylation of the aliphatic chain is observed [25].

The modification of TiO_2 with noble metals has improved the activity of this semiconductor for photodegradation and mineralization of diuron in an aqueous medium. Katsumata et al. impregnated the P25 at different doses of Pt in an oxidized state, stabilizing thermally up to 700°C. They described that 0.2% of Pt in TiO_2 showed the best performance in the photodegradation of diuron in a period of 20 min, and this material is four times more active than pure P25. Nevertheless, 97% of mineralization was reached after 8 h [26].

1.2.2 Methyl parathion

Methyl parathion (O, O-dimethyl O-p-nitrophenyl phosphorothioate) is a white crystalline powder that has a pungent smell like garlic and has low solubility in water (55 mg/L). As insecticide helps to control the biting and sucking of insects in fruit and vegetable crops, it is also applied in the fight against mites, Coleoptera, and caterpillars [27]. Furthermore, methyl parathion is capable to inhibit the action of acetylcholinesterase of nerve tissue, following its metabolic conversion to its corresponding phosphates methyl paraoxon and paraoxon. Organophosphate pesticides are generally regarded as safe for use on crops and animals due to their relatively fast biodegradation, but depend on microbial composition, pH, temperature, and sunlight. This compound can be degraded rapidly by hydrolysis in the presence of sunlight and air [28]. Physical, chemical, and biological methods have been used to minimize the toxic effect generated by this pollutant in water. One of the methods most used for this purpose is heterogeneous photocatalysis with titania, due to its high effectiveness to mineralize organic pollutants in an aqueous medium.

Many reports have been cited in the literature describing the photodegradation of the methyl parathion using TiO_2 with UV light under different conditions. Evgenidou et al. analyzed the photocatalytic behavior of TiO_2 and ZnO in the degradation of methyl parathion in an aqueous medium. They determined that the titania is more effective as a photocatalyst, presenting a higher reaction rate; in addition, this material could complete the mineralization process, without introducing unwanted intermediaries in the reaction [29].

On the other hand, investigations have been carried out involving the modification of TiO_2 to improve its photocatalytic behavior in the degradation of methyl parathion. Senthilnathan and Philip doped the titania with N using different precursors such as triethylamine, ethylamine, urea, and ammonium hydroxide. Their results show that the highest photoactivity is obtained using triethylamine, however, this catalyst when used under UV light did not show a higher performance than P25-TiO_2, but when used with visible irradiation its effectiveness was the best [30].

1.3 TiO_2 doped with rare earth

Rare earth ions have been doped in TiO_2 as a strategy to increase the response of the semiconductor to the visible light region and enhance photocatalytic activity. It was reported in the literature that the optimum level of rare earth doping is less than 2% to hinder the crystal growth of titania during calcination [31]. Also, it is known that the rare earth ions occupy substitutional sites in the titania according to the analyzes carried out by XRD, but in many publications, this statement is contrary, due to the effect of the large ionic radionics of the rare earth ions, which they can only occupy interstitial sites or form aggregates such as oxides or hydroxide at the boundaries of the titania grain by creating Ti-O-RE bonds. The presence of this link generates an imbalance of charges with a positive charge center, which allows adsorbing anions to reach equilibrium. Therefore, the photo-generated holes can be consumed immediately after being transferred to the surface of the titania, whereby the separation of charges is improved; the process of recombination of hollow electron pairs is avoided, and consequently, the photocatalytic activity is favored [16]. Another effect, which inhibits the recombination process and therefore increases the photocatalytic yield of titania, is the formation of the Ti^{3+} species and the oxygen vacancies, both act as photo-generated hole capturers (valence band),

and together they are charged and at the same time, the oxygen of the medium traps the photo-generated electrons (conduction band); and this increases the separation of the photo-generated species. Ti^{3+} is oxidized in the presence of the oxygen present to generate the anion superoxide (O_2^-), which reacts with the photo-generated holes in order to produce hydroxyl radicals (OH *) in an aqueous medium, and thus be able to oxidize any organic compound present in this system [6].

Rare earth dopant ions such as Pr, Ce, Nd, Eu, Sm, Dy, Gd, and La show significant enhancement in dye photodegradation compared with TiO_2 pure, due to the higher adsorption and the *4f* electron transition of rare earth ions. Between 0.5 and 1% wt of doping ions, the best photocatalytic behavior of the doped samples are shown [31]. On the other hand, La, Nd, Sm, Eu, Gd, and Yb as dopants in TiO_2, increase the titania yield and raise the stability of the anatase phase and prevent the segregation of titania. Likewise, these ions play a role in providing a means of concentrating the contaminants to be eliminated on the surface, and consequently, increasing the photocatalytic activity of semiconductor [32]. Recently, La, Nd, Eu, Sm, Gd, Er, Tb, Yb, Pr, and Ce when used as dopants improve the performance of titania, because it increases the absorption capacity of light, to surface and structural modifications, which has allowed the development of catalysts with environmental applications such as the degradation of pollutants in aqueous medium [33].

1.4 Methods of synthesis of TiO_2-RE

There are several methods used for the preparation of TiO_2 doped with rare earth ions, they exist from the most complex and expensive to the simplest and cheapest. These methods vary depending on the final structure that is desired, for example, to obtain thin films or coatings, the following methods are more used: Micro-arc oxidation [34], magnetron sputtering [35], spin coating [36] and dip coating [37]. To prepare powders with defined nanostructures are electrospinning (nanofibers) [38], anodization (nanotubes) [39], microemulsion (spheres) [40], hydrothermal (nanowires) [41], state solid reaction (amorphous) [42], impregnation (amorphous) [43], and sol-gel (different structures) [44]. All of them can be modified, combined with each other or coupled to different energy sources, such as microwaves [45] and ultrasounds [46], to create doped materials with unique photocatalytic properties.

1.4.1 Sol-gel

The sol-gel method has been the most used process for the synthesis of TiO_2 doped with rare earth ions due to the modifications that can make it, at its low cost and easy operation. With this method, crystalline titania can be prepared on a nanometric scale, with a high purity at low temperatures, stoichiometrically controlling the composition when dopants are inserted. This technique is also used for the synthesis of materials with spectral response in the visible region. In this process, the monomers in solution are hydrolyzed and polycondensed to form a polymer network in gel form (during this step, most of the authors report the insertion of doping ions), which contains a liquid phase and a solid phase. After the formation of the gel, the xerogel is formed by removing the solvents in the medium, and then thermal stability is provided by calcining the xerogel at temperatures above 200°C, until obtaining a dense solid with the desired crystalline structure [44].

1.4.2 Impregnation

Wet chemical impregnation is the very simple preparation method to implement to synthesize titania doped with rare earth ions; its process is very easy to perform; it employs mild working conditions and a low energy cost. It provides a uniform distribution of the dopant with the surface of the TiO_2, generating an excellent adhesion between both, which allows controlling the structure, morphology, and particle size simply by modifying conditions such as rotation speed or agitation, contact time, system pH, and nature of solvents. This procedure can be summarized in three simple steps: (1) Place the titania in contact with an aqueous solution containing dissolved dopant precursors for a certain time in constant agitation. (2) Remove excess water in the system, and (3) Activate the material obtained with thermal treatments at elevated temperatures. The variables that mostly influence the preparation of titanium oxide doped with this method are listed below: morphology and structure of TiO_2, amount of dopant material and its disposition with titania, types of solvents used, system pH and type of treatment thermal employee. Under the control of these conditions, this process allows being constantly reproducible [47].

2. Materials and methods

2.1 Characterization techniques of TiO_2-RE

TiO_2 doped with rare earth ions were characterized by XRD, N_2 physisorption, Raman spectroscopy, scanning electron microscopy, and Uv-Vis spectroscopy with diffuse reflectance to describe the electronic, structural, and morphological properties.

2.1.1 UV-Vis (DRS)

UV-Vis spectra were used to estimate the band gap energy (Eg) for each catalyst, if the absorption coefficient (α) is zero, according to Eq. (1). This was performed in a UV-Vis spectrophotometer equipped with an integrating sphere for diffuse reflectance (Varian model Cary 300) using $BaSO_4$ as a reference [48].

$$\propto(\mathbf{hv}) = \mathbf{A}(\mathbf{hv} - \mathbf{Eg})^{\frac{m}{2}} \quad (1)$$

2.1.2 X-ray diffraction (XRD)

A Bruker model D8 advance X-ray diffractometer with anode of Cu Kα radiation (λ = 1.5418 Å) was used. The samples were measure/d in the range of 2θ = 20–70 with a 0.02° step at a rate of 1 s/point at room temperature. To obtain the crystallographic planes of the crystal structures in the samples; they were identified through the library of the Joint Committee on Powder Diffraction Control Standards (JCPDS). The crystal size was calculated in nm (D) of the crystalline phases with the Scherrer Eq. (2) [48]:

$$\mathbf{D} = \frac{\mathbf{K\lambda}}{\boldsymbol{\beta}\ \mathbf{Cos}\ \boldsymbol{\theta}} \quad (2)$$

The percentage of the Rutile phase (%R) was determined using Spurr Eq. (3) [49]:

$$\mathbf{R} = \left(\frac{\mathbf{1}}{\mathbf{1} + \left(\mathbf{0.8} * \frac{\mathbf{I_A}}{\mathbf{I_R}} \right)} \right) * \mathbf{100} \tag{3}$$

2.1.3 BET surface area by N_2 physisorption

Micromeritics ASAP 2020 equipment was used to obtain the specific surface area and pore diameter distribution. Results were calculated from nitrogen adsorption isotherms. Before measuring, the samples were outgassed at 350°C for 2 h.

2.1.4 Raman spectroscopy

Spectra were obtained by a PerkinElmer Spectrum GX NIR-FT Raman spectrometer equipped with a microscope and CCD detector. Spectra were taken at room temperature and using a 5145°A line and argon ion laser (model spectra physics 2020) excited with of 50 mV of energy [50].

2.1.5 Scanning electron microscopy

A scanning electron microscope JEOL model JSM-6010LA was used to identify in detail the surface morphology of the photocatalysts. The analyzer X-Ray Energy Dispersive Spectrometry (XEDS) was used for elemental mapping in materials.

2.2 Synthesis of TiO_2 doped with Ce^{3+}, Pr^{3+}, La^{3+}, Nd^{3+} Sm^{3+}, Eu^{3+}, and Gd^{3+}

Rare earth doping was made using nitrate salts of each element. Water solutions of these salts were prepared to calculate the stoichiometric amount in order to obtain 0.5, 0.3, and 0.1 wt% according to the desired composition of the sample. A mixture of ethanol, water, and salt solution of rare earth was stirred and maintained under reflux at 70°C. Enough NH_4OH was added to the mixture to obtain pH 7. Titanium n-butoxide was added dropwise to this solution, stirring and refluxing was maintained for the period of 24 h until gel formation. The gels were dried in a rotating evaporator at 70°C under vacuum, subsequently, gels were placed into an oven at 120°C during 12 h. Samples were calcined at 500 and 800°C during 4 h with a heating ramp of 2°C/min. The same procedure was carried out but without adding the precursors of the rare earth ions to obtain the pure titania [51].

2.3 Photocatalytic reactions

The photocatalytic activity was evaluated following the degradation with respect to the time of two aqueous solutions of diuron (249 nm) and another of methyl parathion (275 nm), respectively. The conversion was measured by UV-Vis spectroscopy of the peaks of maximum absorbance for each pollutant. For this process was used a borosilicate Ace photocatalytic reactor that was enabled with a quartz recirculation system to maintain a constant reaction temperature of 20°C. The reaction total volume was 300 mL, employing 0.5 g/L as a catalyst concentration, constant stir and flow of 60 ml/seg atmospheric O_2. The total reaction time was 300 min. The reactions were carried out under sunlight, and the fully equipped reactor was placed on the roof of the laboratory building in full sunlight from 10:00 a.m. to 3:00 p.m.

3. Results and discussion

3.1 Results of characterization of TiO_2 doped with Ce^{3+}, Pr^{3+}, La^{3+}, Nd^{3+} Sm^{3+}, Eu^{3+}, and Gd^{3+}

The diffuse reflectance spectra in the materials describe a change in the absorption band toward wavelengths corresponding to the visible region in all materials calcined at 500 (**Figure 1a** and **b**) and 800°C (**Figure 1c**). Due to the elimination of impurities, organic material and hydroxylated groups from the precursors in the synthesis, which use NH_4OH as hydrolysis catalyst and NO_3^- ions as precursors of the dopants, so it is expected that N atoms have been incorporated and eliminated by the effect of thermal treatment, causing a shift of the absorption bands toward longer [52], at the same time, this caused the production of oxygen vacancies, which also produced the same effect [53].The spectra of **Figure 1a** present a greater absorption toward the visible for the solids doped with Pr 0.3%, Nd 0.1%, and 0.3%, however, for **Figure 1b**, the sample with Sm 0.3% is the one that greater absorption shows and in **Figure 1c**, the materials doped with Sm 0.3% and Gd 0.3% have this same behavior. In **Figure 1c**, it is evident that materials calcined at 800°C have a better absorption toward the visible, compared with those calcined at 500°C, due to the presence of the rutile crystalline phase, due to the increase in temperature in the thermal treatment, which consequently produces a decrease in the value of the band gap energy, the same happens with the sample of pure TiO_2 treated at 800°C. All doped materials already treated at 500 and 800°C show greater absorption than pure TiO_2. The values of the Eg are shown in **Table 1**, here, it is observed that most of the samples treated at 500°C have a value ranging between 3.02 and 3.16 eV, including doped materials and pure TiO_2, however, only the sample doped with Pr 0.3% decreased this value considerably (2.88 eV). With respect to the solids treated at 800°C, the photocatalyst doped with Eu 0.3% obtained the lowest value of Eg compared with all materials (2.5 eV).

Figure 2 shows the diffractograms corresponding to pure TiO_2 and TiO_2-doped with rare earth ions, all thermally treated at 500°C. Only the anatase crystalline phase was detected in these samples presenting a low crystallinity (slightly amorphous), which is observed in the morphology by SEM in **Figure 6a**.

The results in Table 1 show that an average crystal size is 8.14 nm for the TiO2 calcined at 500°C when introducing the dopants at the same treatment temperature the value decreases until 7.00 nm for the case of the catalyst with Sm 0.5%. In most of the samples doped to increase their content in the titania, the average crystal size decreases; this is due to the separation of the dopant in the limits of the crystal, which prevents its growth by not being able to be in contact with other crystals to grow by coalescence [54], except for the Nd and Pr, where their values are fixed

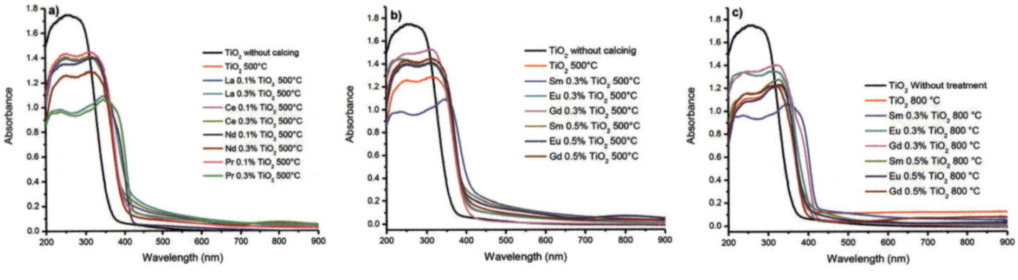

Figure 1.
UV-Vis diffuse reflectance spectra of TiO_2 without treatment and TiO_2 (a) doped to 0.1 and 0.3% with La, Ce, Nd, and Pr calcinated at 500°C; (b) doped to 0.3 and 0.5% with Sm, Eu, and Gd calcinated at 500°C; (c) doped to 0.3 and 0.5% with Sm, Eu, and Gd calcinated at 800°C.

Samples	Specific Area Bet (m^2/g)	Pore Diameter (nm)	Average Crystal Size (nm)	% Mixture Phase Crystalline	Band Gap Energy (ev)
TiO_2 500 °C	78.71	7.91	9.91	100% A	3.02
TiO_2La0.1 500°C	90.10	9.50	8.41	100% A	3.07
TiO_2La0.3 500°C	119.00	9.50	7.21	100% A	3.10
TiO_2Ce0.1 500°C	98.01	9.50	8.41	100% A	3.00
TiO_2Ce0.3 500°C	109.04	9.50	9.01	100% A	3.10
TiO_2Nd0.1 500°C	99.00	7.81	8.41	100% A	2.98
TiO_2Nd0.3 500°C	111.06	7.70	8.41	100% A	3.00
TiO_2Pr0.1 500°C	99.09	7.80	8.41	100% A	3.05
TiO_2Pr0.3 500°C	112.01	7.70	8.41	100% A	2.88
TiO_2Sm0.1 500°C	105.70	7.70	8.25	100% A	3.16
TiO_2Sm0.3 500°C	109.50	9.40	7.07	100% A	3.10
TiO_2Sm0.5 500°C	95.91	8.42	7.00	100% A	3.09
TiO_2Eu0.1 500°C	101.80	9.42	9.91	100% A	3.09
TiO_2Eu0.3 500°C	100.50	9.49	8.25	100% A	3.09
TiO_2Eu0.5 500°C	104.70	7.87	8.20	100% A	3.08
TiO_2Gd0.1 500°C	113.32	7.6	8.25	100% A	3.23
TiO_2Gd0.3 500°C	93.91	9.43	8.25	100% A	3.11
TiO_2Gd0.5 500°C	92.48	8.14	8.21	100% A	3.07
TiO_2 800 °C	16.70	32.05	133.66	100% R	2.90
TiO_2Sm0.3 800°C	24.73	30.71	32.10 A 140.81 R	46.40% A 53.60% R	2.80
TiO_2Eu0.3 800°C	8.48	35.23	34.43 A 132.67 R	38.65% A 61.35% R	2.50
TiO_2Gd0.3 800°C	26.64	30.21	31.18 A 126.67 R	40.22% A 59.78% R	2.88

Table 1.
Results of characterization techniques applied to TiO_2 and doped TiO_2: N_2 physisorption, XRD and UV-Vis spectroscopy with RD.

and possibly with these ions to increase their content will no longer reduce their average crystal size. As mentioned above, the presence of rare earth ions inhibits the complete transformation of phases due to temperature; for our case, it happens at 800°C where catalysts stabilized at this temperature are present and clearly show mixtures of crystalline phases (anatase-rutile), and this is observed in **Figure 3**. A commercial sample of titania (P25-TiO_2) was compared with the materials calcined at the highest temperature with this technique. It was observed that the materials doped with rare earth ions showed a greater intensity in the peaks corresponding to the rutile phase, which describes a greater abundance of this phase and is less than 70% because this is the approximate percentage of rutile phase reported for this solid [55]. This can be corroborated with the results of **Table 1**, where the percentage of crystalline phase is described; here, the samples doped and calcined at 800°C indicate mixtures of phases with close proportions for Eu and Gd (40% A-60% R); however, for the Sm, the proportion percentage was almost equal (47% A-53% R). The above can also be confirmed according to the Raman spectra shown in

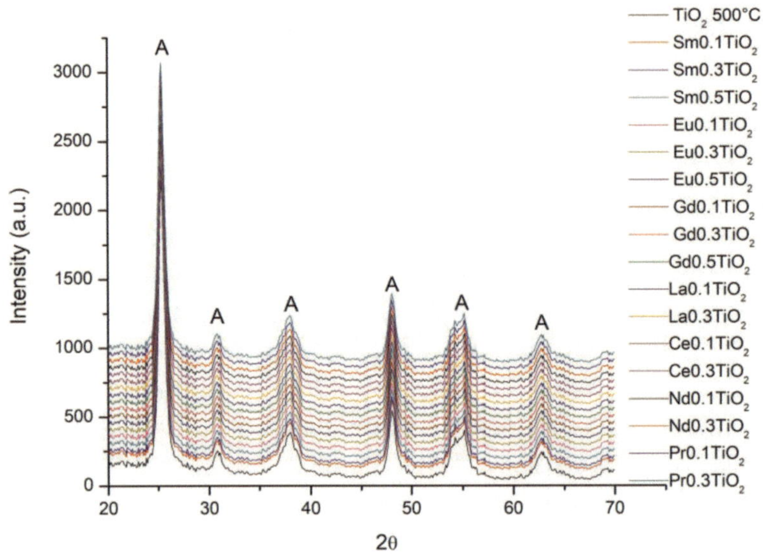

Figure 2.
XRD patterns of pure TiO_2 and doped TiO_2 with rare earth thermally treated to 500°C.

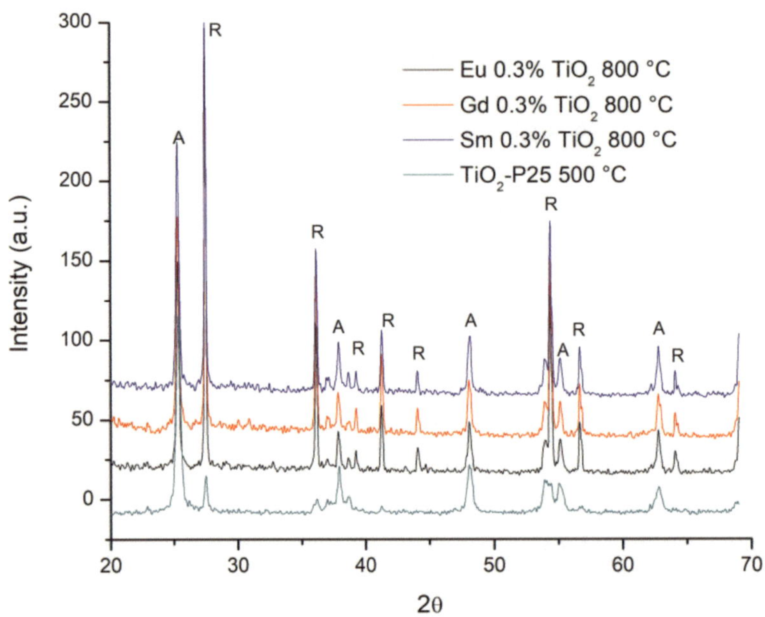

Figure 3.
XRD patterns of P25-TiO_2 and doped TiO_2 with rare earth thermally treated to 800°C.

Figure 4, in which the peaks corresponding to the anatase and rutile phases respectively are described. The phase mixtures at high temperatures can be explained due to the connection between the Ti^{4+} (octahedral) and RE^{3+} (tetrahedral) ions, where the Ti^{4+} ions replace the surface RE^{3+} ions in the network the rare earth oxide to form sites tetragonal of Ti, the interaction between atoms of Ti^{4+} octahedral and Ti^{4+} tetrahedral prevents the transformation of phases in the thermal treatment [56]. The average crystal size for these samples is also found in **Table 1** and calculated individually for each crystalline phase. With respect to the anatase phase, the highest value was presented by the sample doped with Eu (34.43 nm), which is more than three times greater than the sample of pure titania at 500°C. However, the material doped with Sm obtained the highest average crystal size for the rutile phase (140.81 nm).

Figure 5 shows the adsorption-desorption isotherms of the materials calcined at 500°C. It is noteworthy that the incorporation of the dopant into TiO_2 generates greater physical adsorption by increasing the relative pressure, explained by a possible uniform surface dispersion of the dopant, which demonstrates the increase in the specific area with respect to pure TiO_2 and the capacity of the dopant. Rare

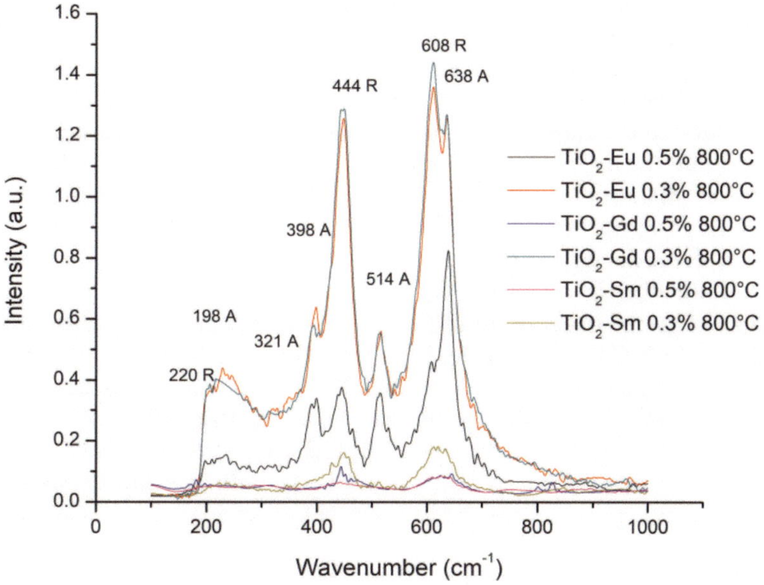

Figure 4.
Raman spectra of TiO_2 doped with Sm, Eu, and Gd, calcinated at 800°C.

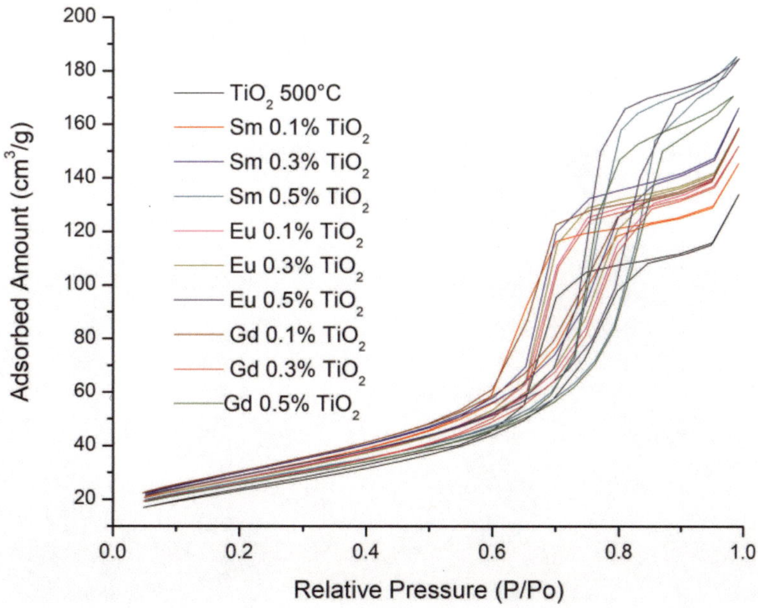

Figure 5.
Adsorption-desorption isotherms of TiO_2 and doped TiO_2 with Sm, Eu, and Gd calcinated al 500°C.

earth ions to form complexes with several Lewis bases. It is observed that all the isotherms have a type IV behavior according to the IUPAC classification, which is a characteristic of mesoporous solids and has multiple layer adsorption mechanism, with a hysteresis loop of type A according to the same organism, which indicates the description of mesoporous solids with capillaries in tubular form and ink cans; these samples have a desorption of similar geometric shape although their adsorption varied with respect to the metal and the amount of dopant.

The specific area values of the pure, doped, and calcined catalysts at 500 and 800°C are compiled in **Table 1**. As expected, the presence of the dopants in the titania increases the specific area for all materials treated at 500°C, which previously had already been described with other materials [51]. The value of this parameter was between 90.10 and 119.50 m^2/g, the sample doped with Ce 0.3% presented the highest value, and this increase can be attributed to (1) the high dispersion that had the rare earth ions and this can be seen in **Figure 6b** with the image describing the elemental dispersion of Sm at 0.3% in titania, which also manifests with all doping ions, (2) to the impediment of rare earth ions to enter the lattice titania due to the large size of its ions, (3) the low amount of dopant that was used and (4) the reduction in the size of the crystal. This confirms that the rare earth ions inhibit the sintering of TiO_2 [57].

The thermal transformation to rutile in the titania decreases considerably the specific area to 800°C and therefore, increases its crystallinity and the sintering process, but when inserting Sm and Eu to 0.3%, the area increases due to the formation of mixtures of crystalline phases, distorting the surface of the titania due to the presence of dopants. The titania catalyst doped with Eu at 0.3% and calcined at 800°C does not increase its specific area; it is the only one that shows a value below pure TiO_2; this possibly at the low percentage ratio that it has anatase phase.

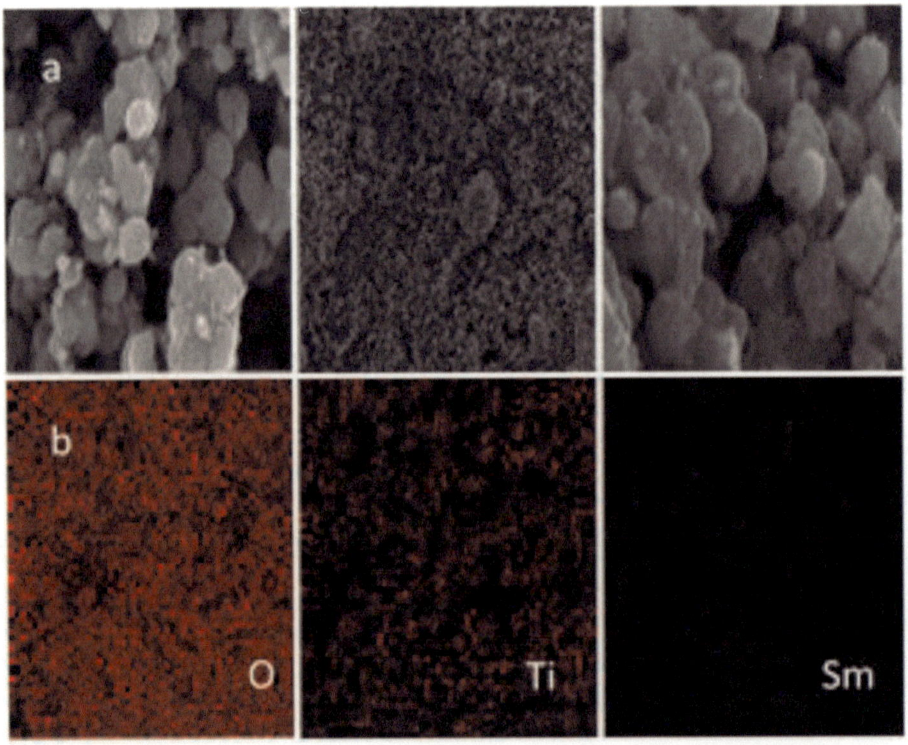

Figure 6.
Images obtained by SEM to the Sm 0.3 TiO_2 500°C: (a) morphology and (b) elemental mapping.

The pore diameter showed an increase with respect to pure TiO_2 with samples doped with La, Ce at different contents, with Eu at 0.1 and 0.3%, with Sm 0.3 and 0.5%, and with Gd 0.3 and 0.5% all calcined at 500°C. This increase is attributed to the fusion of small pores present in the anatase phase to form large pores or stacked cavities. However, the reduction in the value of this parameter is explained by the possible blocking of porous cavities by the dopants. Similarly, at 800°C, the materials showed similar behavior, but to these conditions, the increase in the pore diameter is due to pore coalescence during calcination [51].

3.2 Results of photocatalytic tests of diuron using TiO_2-RE

The photocatalytic activity under sunlight in the photodegradation of diuron is described in **Figure** 7. All materials doped and calcined at 500°C obtained a higher yield than photolysis, pure TiO_2 and P25-TiO_2 (**Figure 7a**). This behavior is directly related to the high dispersion of dopants, which superficially modifies the titania, considerably increasing its specific area. The materials doped with La and Ce 0.1%, respectively, are the most active of the series when starting the dopant content, show an increase in the volume of pores, which indicates that the dispersion of these dopants do not obstruct the porous cavities, generating a porous material that has greater contact with the polluting solution, the opposite occurs with the rest of

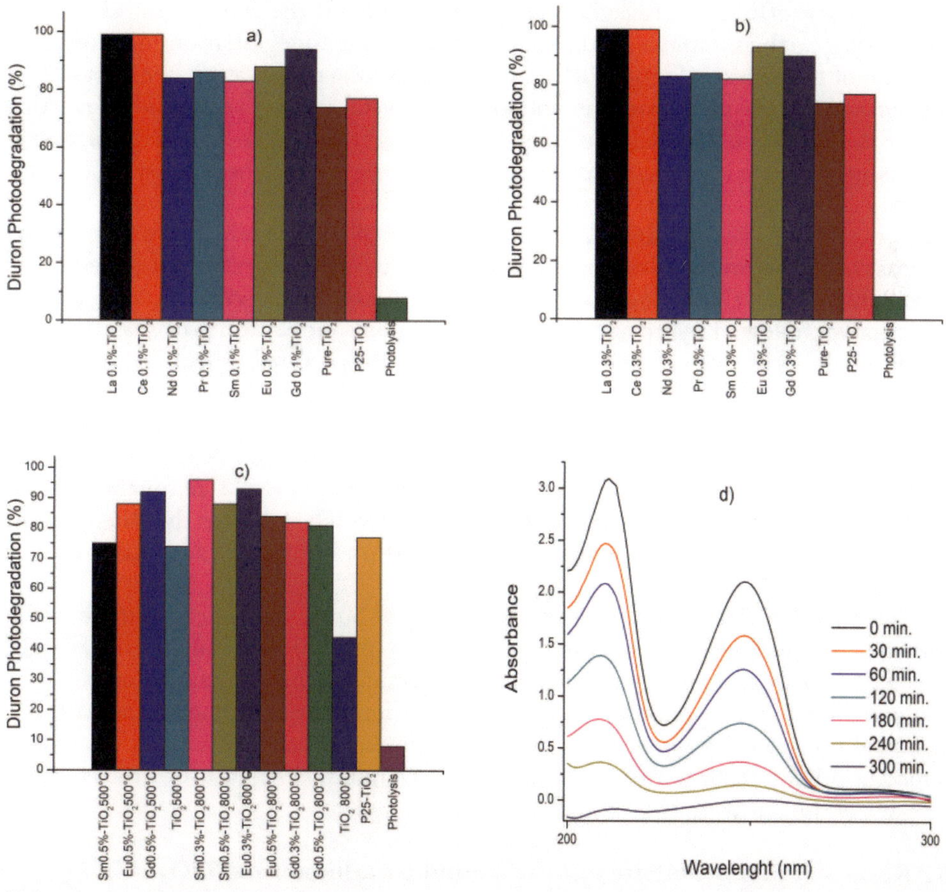

Figure 7.
Photocatalytic degradation of diuron using solar light: (a) doped photocatalyst to 0.1% at 500°C. (b) Doped photocatalyst to 0.3% at 500°C. (c) Doped photocatalyst to 0.5% at 500°C and doped photocatalyst to 0.3–0.5% at 800°C. (d) Solar photodegradation of diuron with Ce 0.1% TiO_2 500°C.

the doping ions, which upon insertion decrease the pore volume and its photoactivity decreases. For the sample with Eu, although it also increases its pore volume with respect to pure titania, its activity declines considerably and has the worst photocatalytic performance, when contrasting this phenomenon with the average particle sizes. It was found that this material maintains the same particle size; therefore, it can be attributed that photocatalytic activity in this material depends on the average particle size.

When increasing the content to 0.3% of the dopants in the titania, the materials doped with Eu and Gd do not increase the specific area with respect to the photocatalysts, in which the samples doped only with La, Ce, and Sm increase their photocatalytic performance, as seen in **Figure 7b**. The first two conserve the same pore diameter, showing a good dispersion by augmenting the dopant content that consequently raises the specific area, without blocking pores as reported in other investigations [19]. In the sample with Sm at this content, the dispersion improves remarkably with respect to 0.1%, which was observed in **Figure 6b** in the elemental mapping by EDS, which increases the pore diameter in the material. The tendency in the size of crystal when increasing concentration of dopant describes a reduction and conservation in values of this parameter, which already has been previously reported [54]; only the material doped with Ce increases this value, but not in a way significant and still showing a smaller size with respect to pure titania.

In **Figure 7c**, it is observed that the increase dopant content to 0.5% decrease photocatalytic efficiency; this gives us an idea of what is the content and the ideal ions of rare earth as dopants in the titania for solar photodegradation in aqueous medium of the diuron (**Figure 7d**), as in other investigations, at higher dopant concentrations, the pore diameter and average crystal size decrease [58]. Nevertheless, if the content of doping ions is excessively high, the recombination process of the photo-generated becomes easier, which led to the lower photocatalytic activity of titania. The photoactivity of some samples calcined at 800°C, presented a better efficiency than pure TiO_2 and P25-TiO_2; the trend of the specific area describes an increase with the presence of dopants, but with Eu, this parameter decreased considerably to a value below pure titania. The photocatalyst with Sm 0.3% was the most active, followed by Eu and Gd at this treatment temperature; this is attributed mainly to the presence of mixtures of crystalline phases (anatase-rutile), which incorporates Sm; the ratio of this mixture indicates a higher content of anatase (about 46%), as described above. This phase is more active than rutile and has a higher anatase ratio than that reported for P25-TiO_2 (30% approx.); therefore, the performance is better. However, this same sample has a crystal size for the greater rutile phase compared to the rest of the materials, which indicates that the growth of the rutile crystals is directly proportional to the photocatalytic activity of the materials.

The materials doped and thermally stabilized at 500°C more active in the previous reaction, were selected to evaluate them photocatalytically under sunlight in the degradation of an organophosphorus insecticide (methyl parathion), to analyze the effect that photocatalysts have on different aqueous pollutants with different functional groups (phenylurea and thiophosphate). Although materials doped at 0.1 and 0.3% with La and Ce obtained similar yields, those with the least amount of dopant were chosen.

3.3 Results of photocatalytic tests of methyl parathion using TiO_2-RE

The photocatalytic behavior of the catalysts chosen for this test is described in **Figure 8a**; in **Figure 8b**, the photodegradation of methyl parathion with respect to time is shown in the most photoactive material of this series of materials doped and

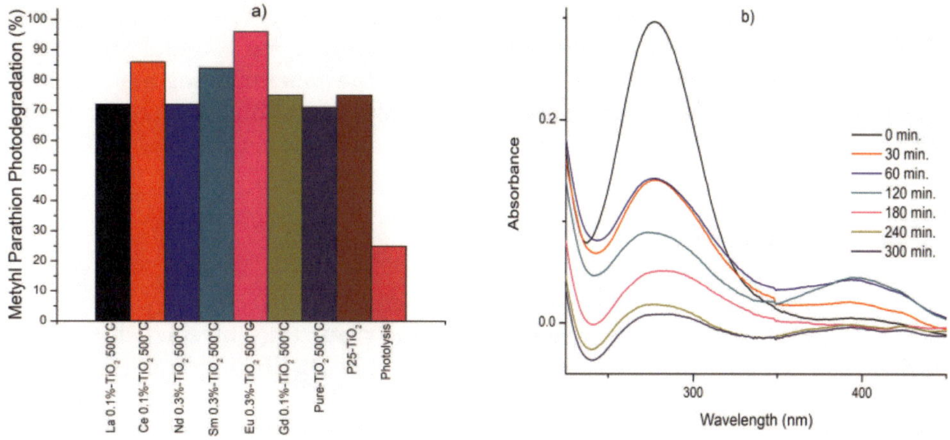

Figure 8.
Photocatalytic degradation of methyl parathion using solar light: (a) La 0.1% TiO_2, Ce 0.1% TiO_2, Nd 0.3% TiO_2, Sm 0.3% TiO_2, Eu 0.3% TiO_2, Gd 0.1% TiO_2, pure TiO_2, P25-TiO_2 and photolysis. (b) Solar photodegradation of methyl parathion with Eu 0.3% TiO_2 500°C.

calcined at 500°C. A different trend is observed when comparing the activity of the materials with the previous model molecule (diuron). The conversion by photolysis without catalyst was around 25%; however, conversions are reached above 95% with TiO_2 doped with Eu at 0.3% and with the other catalysts, the conversions exceed 70%. Only samples doped with Eu 0.3%, Ce 0.1%, and Sm 0.3% were more active than P25-titania. The value of the average diameter of pores in these materials increased and was compared with the pure titania; this increment also happened with the doped catalyst with La, but as its specific area was the lowest, this reduced its photoactivity. This seems to indicate that the presence of pores and specific area large, allows a better diffusion between the polluting solution and the photocatalyst to increase the photoactivity of TiO_2 doped with rare earth ions. With respect to the average crystal size, all materials decrease this value due to the presence of rare earth ions, which inhibit the growth of crystals in the process of synthesis and thermal stabilization. Finally, it can be stated that there is an affinity between the doped ions in the titania and the main functional groups of the molecules used for the photodegradation under sunlight. With phenylurea (diuron), solar photodegradation was more pronounced with the materials doped with La and Ce, meanwhile, for the thiophosphate (methyl parathion), the process of solar photo-oxidation had a greater affinity for the Eu.

4. Conclusions

The rare earth doping ions improve the textural, structural, electronic, and photocatalytic properties in TiO_2. Due to the method of preparation and the treatment temperature, possibly the presence of N and the elimination of impurities produced a change in the absorption bands, which allows the titania to have a better photocatalytic behavior under sunlight. At 500°C, the materials present 100% of the anatase crystalline structure, the ideal amount of dopant was 0.1% and the most active rare earth ions were La and Ce in the diuron solar photodegradation. The increase in temperature of thermal treatment (800°C) showed the presence of mixtures of crystalline phases, which have a greater abundance of anatase, compared to P25-TiO_2, where the catalyst doped with Sm 0.3% obtained the best performance. Photocatalysts treated at 500°C with greater activity in diuron degradation were chosen to evaluate their solar photoactivity with a second pesticide (methyl parathion). Under these conditions, an affinity was found for the dopant ions in titania and the functional groups of the contaminating molecules (phenylurea and thiophosphate). Solar photodegradation of diuron was more effective with La and Ce, while for methyl parathion it was Eu at 0.3%.

Acknowledgements

The authors thank the National Council for Science and Technology (CONACYT) for financing the project 132648 and thank the Universidad Juárez Autónoma de Tabasco for PFCE-DACB Project and PRODEP Program.

Conflict of interest

The authors declare no conflicts of interest.

Author details

Juan C. Arévalo Pérez*, José Gilberto Torres Torres, Durvel de la Cruz Romero, Hermicenda Perez-Vidal, Maria Antonia Lunagomez Rocha, Ignacio Cuauhtémoc López, Adrian Cervantes Uribe and Zenaida Guerra Que
Laboratory of Catalytic Nanomaterials Applied to the Development of Energy Sources and Environmental Remediation, Applied Science and Technology Research Center of Tabasco (CICTAT), Juarez Autonomous University of Tabasco, DACB, Cunduacan, Tabasco, México

*Address all correspondence to: juan.carlos.arevalo.jcap@gmail.com

References

[1] Cerejeira MJ, Viana P, Batista S, Pereira T, Silva E, Val MJ, et al. Pesticides in Portuguese surface and ground waters. Water Research. 2003; **37**:1055-1063. DOI: 10.1016/S0043-1354(01)00462-6

[2] Parimi S, Meinke LJ, Wade French B, Chandler LD, Siegfried BD. Stability and persistence of aldrin and methyl-parathion resistance in western corn rootworm populations (Coleoptera: Chrysomelidae). Crop Protection. 2006; **25**:269-274. DOI: 10.1016/j.cropro.2005.04.017

[3] Zhang J, Zheng Z, Zhao T, Zhao Y, Wang L, Zhong Y, et al. Radiation-induced reduction of diuron by gamma-ray irradiation. Journal of Hazardous Materials. 2008;**151**:465-472

[4] Taylor P, Krishna KR, Philip L. Biodegradation of lindane, methyl parathion and carbofuran by various enriched bacterial isolates. Journal of Environmental Science and Health, Part B. 2008:37-41. DOI: 10.1080/03601230701795155

[5] Kim TS, Kim JK, Choi K, Stenstrom MK, Zoh KD. Degradation mechanism and the toxicity assessment in TiO_2 photocatalysis and photolysis of parathion. Chemosphere. 2006; **62**:926-933. DOI: 10.1016/j.jss.2005.06.046

[6] Saqib N, Adnan R, Shah I. A mini-review on rare earth metal-doped TiO_2 for photocatalytic remediation of wastewater. Environmental Science and Pollution Research. 2016:15941-15951. DOI: 10.1007/s11356-016-6984-7

[7] Carraway ER, Hoffman AJ, Hoffmann MR. Photocatalytic oxidation of organic-acids on quantum-sized semiconductor colloids. Environmental Science & Technology. 1994;**28**:786-793. DOI: 10.1021/es00054a007

[8] Choi H, Antoniou MG, Pelaez M, De La Cruz AA, Shoemaker JA, Dionysiou DD. Mesoporous nitrogen-doped TiO_2 for the photocatalytic destruction of the cyanobacterial toxin microcystin-LR under visible light irradiation. Environmental Science and Technology. 2007;**41**:7530-7535. DOI: 10.1021/es0709122

[9] Wilke K, Breuer HD. The influence of transition metal doping on the physical and photocatalytic properties of titania. Journal of Photochemistry and Photobiology A: Chemistry. 1999; **121**:49-53. DOI: 10.1016/S1010-6030(98)00452-3

[10] Pan X, Xu YJ. Defect-mediated growth of noble-metal (Ag, Pt, and Pd) nanoparticles on TiO_2 with oxygen vacancies for photocatalytic redox reactions under visible light. Journal of Physical Chemistry C. 2013;**117**: 17996-18005. DOI: 10.1021/jp4064802

[11] Ma Y, Zhang J, Tian B, Chen F, Wang L. Synthesis and characterization of thermally stable Sm, N co-doped TiO_2 with highly visible light activity. Journal of Hazardous Materials. 2010;**182**: 386-393. DOI: 10.1016/j.jhazmat.2010.06.045

[12] Chen FY, Cao WK, He SY, Wang BH, Zhang YM. Synthesis, characterization and thermochemistry properties of RE(III) and 2-oxo-propionic acid salicyloyl hydrazone complexes. Acta Physico-Chimica Sinica. 2006;**22**:280-285. DOI: 10.1016/S1872-1508(06)60003-X

[13] Mazierski P, Mikolajczyk A, Bajorowicz B, Malankowska A, Zaleska-Medynska A, Nadolna J. The role of lanthanides in TiO_2 based photocatalysis: A review. Applied Catalysis B: Environmental. 2018;**233**: 301-317. DOI: 10.1016/j.apcatb.2018.04.019

[14] Zhang W, Yang S, Li J, Gao W, Deng Y, Dong W, et al. Applied catalysis B: Environmental visible to ultraviolet Upconvertion: Energy transfer, material matrix, and synthesis strategies. Applied Catalysis B, Environmental. 2017;**206**: 89-103. DOI: 10.1016/j.apcatb.2017.01.023

[15] Lin J, Yu JC. An investigation on photocatalytic activities of mixed TiO_2 rare earth oxides for the oxidation of acetone in air. Journal of Photochemistry and Photobiology A: Chemistry. 1998;**116**:63-67. DOI: 10.1016/S1010-6030(98)00289-5

[16] Yu Y, Chen G, Zhou Y, Han Z. Recent advances in rare-earth elements modification of inorganic semiconductor-based photocatalysts for efficient solar energy conversion: A review. Journal of Rare Earths. 2015;**33**: 453-462. DOI: 10.1016/S1002-0721(14)60440-3

[17] Zhu J, Xie J, Chen M, Jiang D, Wu D. Low temperature synthesis of anatase rare earth doped titania-silica photocatalyst and its photocatalytic activity under solar-light. Colloids and Surfaces A: Physicochemical and Engineering Aspects. 2010;**355**:178-182. DOI: 10.1016/j.colsurfa.2009.12.016

[18] Lin L, Chai Y, Yang Y, Wang X, He D, Tang Q, et al. Hierarchical Gd-La codoped TiO_2 microspheres as robust photocatalysts. International Journal of Hydrogen Energy. 2013;**38**:2634-2640. DOI: 10.1016/j.ijhydene.2012.11.100

[19] El-Bahy ZM, Ismail AA, Mohamed RM. Enhancement of titania by doping rare earth for photodegradation of organic dye (direct blue). Journal of Hazardous Materials. 2009;**166**:138-143. DOI: 10.1016/j.jhazmat.2008.11.022

[20] Baiju KV, Periyat P, Shajesh P, Wunderlich W, Manjumol KA, Smitha VS, et al. Mesoporous gadolinium doped titania photocatalyst through an aqueous sol-gel method. Journal of Alloys and Compounds. 2010;**505**: 194-200. DOI: 10.1016/j.jallcom.2010.06.028

[21] Yan QZ, Su XT, Huang ZY, Ge CC. Sol-gel auto-igniting synthesis and structural property of cerium-doped titanium dioxide nanosized powders. Journal of the European Ceramic Society. 2006;**26**:915-921. DOI: 10.1016/j.jeurceramsoc.2004.11.017

[22] Liang CH, Li FB, Liu CS, Lü JL, Wang XG. The enhancement of adsorption and photocatalytic activity of rare earth ions doped TiO_2 for the degradation of Orange I. Dyes and Pigments. 2008;**76**:477-484. DOI: 10.1016/j.dyepig.2006.10.006

[23] Castillo MA, Felis N, Aragón P, Cuesta G, Sabater C. Biodegradation of the herbicide diuron by streptomycetes isolated from soil. International Biodeterioration and Biodegradation. 2006;**58**:196-202. DOI: 10.1016/j.ibiod.2006.06.020

[24] Giacomazzi S, Cochet N. Environmental impact of diuron transformation: A review. Chemosphere. 2004;**56**:1021-1032. DOI: 10.1016/j.chemosphere.2004.04.061

[25] Macounová K, Krýsová H, Ludvík J, Jirkovský J. Kinetics of photocatalytic degradation of diuron in aqueous colloidal solutions of Q-TiO_2. Journal of Photochemistry and Photobiology A: Chemistry. 2003;**156**:273-282. DOI: 10.1016/S1010-6030(02)00091-6

[26] Katsumata H, Sada M, Nakaoka Y, Kaneco S, Suzuki T, Ohta K. Photocatalytic degradation of diuron in aqueous solution by platinized TiO_2. Journal of Hazardous Materials. 2009; **171**:1081-1087. DOI: 10.1016/j.jhazmat.2009.06.110

[27] Pino N, Peñuela G. Simultaneous degradation of the pesticides methyl

parathion and chlorpyrifos by an isolated bacterial consortium from a contaminated site. International Biodeterioration and Biodegradation. 2011;**65**:827-831. DOI: 10.1016/j.ibiod.2011.06.001

[28] Ragnarsdottir K. Environmental fate and toxicology of organophosphate pesticides. Journal of the Geological Society. 2000;**157**:859-876. DOI: 10.1007/978-3-319-03777-6_3

[29] Evgenidou E, Konstantinou I, Fytianos K, Poulios I, Albanis T. Photocatalytic oxidation of methyl parathion over TiO_2 and ZnO suspensions. Catalysis Today. 2007;**124**: 156-162. DOI: 10.1016/j.cattod.2007.03.033

[30] Senthilnathan J, Philip L. Photodegradation of methyl parathion and dichlorvos from drinking water with N-doped TiO_2 under solar radiation. Chemical Engineering Journal. 2011;**172**:678-688. DOI: 10.1016/j.cej.2011.06.035

[31] Zhang T, Oyama T, Horikoshi S, Zhao J, Hidaka H, Serpone N. Significant effect of lanthanide doping on the texture and properties of nanocrystalline mesoporous TiO_2. Journal of Solid State Chemistry. 2004; **177**:3490-3498. DOI: 10.1016/j.jssc.2004.05.026

[32] Bossmann SH, Braun AM. Lanthanide oxide-doped titanium dioxide photocatalysts: Novel photocatalysts for the enhanced degradation of p-chlorophenoxyacetic acid. Environmental Science Technology. 2001;**35**:1544-1549. DOI: 10.1021/es001613e

[33] Yurtsever HA, Çiftçioğlu M. The effect of rare earth element doping on the microstructural evolution of sol-gel titania powders. Journal of Alloys and Compounds. 2016;**695**:1336-1353. DOI: 10.1016/j.jallcom.2016.10.275

[34] Di S, Guo Y, Lv H, Yu J, Li Z. Microstructure and properties of rare earth CeO_2-doped TiO_2 nanostructured composite coatings through micro-arc oxidation. Ceramics International. 2015; **41**:6178-6186. DOI: 10.1016/j.ceramint.2014.12.134

[35] Wojcieszak D. Analysis of Eu-effect on stabilization of the TiO_2-anatase structure in high temperature and photoluminescence efficiency for the coatings as-deposited in magnetron sputtering process. Applied Surface Science. 2017;**421**:128-133. DOI: 10.1016/j.apsusc.2017.01.040

[36] Labreche F, Berbadj A, Brihi N, Karima R, Jamoussi B. Green photoluminescence, structural and optical properties of Nd-TiO_2 thin films. Optik. 2018;**172**:63-71. DOI: 10.1016/j.ijleo.2018.06.131

[37] Borlaf M, Colomer MT, Moreno R, Ortiz AL. Rare earth-doped TiO_2 nanocrystalline thin films: Preparation and thermal stability. Journal of the European Ceramic Society. 2014;**34**: 4457-4462. DOI: 10.1016/j.jeurceramsoc.2014.07.008

[38] Hassan MS, Amna T, Yang O, Kim H, Khil M. TiO_2 nanofibers doped with rare earth elements and their photocatalytic activity. Ceramics International. 2012;**38**:5925-5930. DOI: 10.1016/j.ceramint.2012.04.043

[39] Mazierski P, Lisowski W, Grzyb T, Winiarski MJ, Klimczuk T, Mikołajczyk A, et al. Enhanced photocatalytic properties of lanthanide-TiO_2 nanotubes: An experimental and theoretical study. Applied Catalysis B, Environmental. 2016;**205**:376-385. DOI: 10.1016/j.apcatb.2016.12.044

[40] Jian Z, Pu Y, Fang J, Ye Z. Microemulsion synthesis of nanosized TiO_2 particles doping with rare-earth and their photocatalytic activity. Photochemistry and Photobiology.

2010;**86**:1016-1021. DOI: 10.1111/j.1751-1097.2010.00773.x

[41] Bandi VR, Raghavan CM, Grandhe BK, Kim SS, Jang K, Shin DS, et al. Synthesis, structural and optical properties of pure and rare-earth ion doped TiO_2 nanowire arrays by a facile hydrothermal technique. Thin Solid Films. 2013;**547**:207-211. DOI: 10.1016/j.tsf.2013.03.039

[42] Tobaldi DM, Pullar RC, Škapin AS, Seabra MP, Labrincha JA. Visible light activated photocatalytic behaviour of rare earth modified commercial TiO_2. Materials Research Bulletin. 2014;**50**:183-190. DOI: 10.1016/j.materresbull.2013.10.033

[43] Zhang J, Wu W, Yan S, Chu G, Zhao S, Wang X, et al. Enhanced photocatalytic activity for the degradation of rhodamine B by TiO_2 modified with Gd_2O_3 calcined at high temperature. Applied Surface Science. 2015;**344**:249-256. DOI: 10.1016/j.apsusc.2015.03.078

[44] Akpan UG, Hameed BH. The advancements in sol-gel method of doped-TiO_2 photocatalysts. Applied Catalysis A: General. 2010;**375**:1-11. DOI: 10.1016/j.apcata.2009.12.023

[45] Dominguez RD, Alarcón-Flores G, Aguilar-Frutis M, Sánchez-Alarcón RI, Falcony C, Dorantes-Rosales HJ, et al. Effect on the stabilization of the anatase phase and luminescent properties of samarium-doped TiO_2 nanocrystals prepared by microwave irradiation. Journal of Alloys and Compounds. 2016;**687**:121-129. DOI: 10.1016/j.jallcom.2016.06.083

[46] Su W, Chen J, Wu L, Wang X, Wang X, Fu X. Visible light photocatalysis on praseodymium (III)-nitrate modified TiO_2 prepared by an ultrasound method. Applied Catalysis B: Environmental. 2008;**77**:264-271. DOI: 10.1016/j.apcatb.2007.04.015

[47] Zhang W, Zou L, Wang L. Photocatalytic TiO_2/adsorbent nanocomposites prepared via wet chemical impregnation for wastewater treatment: A review. Applied Catalysis A: General. 2009;**371**:1-9. DOI: 10.1016/j.apcata.2009.09.038

[48] Arévalo Pérez JC, Torres JG, Cervantes Uribe A, Pérez Vidal H, Cordero García A, Izquierdo Colorado A, et al. Use of multivariable analysis (ANOVA) to compare irradiation sources on diuron destruction by photocatalysis using TiO_2-P25 impregnated with Sm^{3+}, Eu^{3+} and Gd^{3+}. Journal of Chemical Engineering & Process Technology. 2018;**9**:1-8. DOI: 10.4172/2157-7048.1000374

[49] Spurr RA, Myers H. Quantitative analysis of anatase-rutile mixtures with an X-ray diffractometer. Analytical Chemistry. 1957;**29**:760-762. DOI: 10.1021/ac60125a006

[50] De la Cruz D, Arévalo JC, Torres G, Bautista Margulis RG, Ornelas C, Aguilar-elguézabal A. TiO_2 doped with Sm^{3+} by sol–gel : Synthesis, characterization and photocatalytic activity of diuron under solar light. Catalysis Today. 2011;**166**:152-158. DOI: 10.1016/j.cattod.2010.08.023

[51] De la Cruz Romero D, Torres GT, Arévalo JC, Gomez R, Aguilar-Elguezabal A. Synthesis and characterization of TiO_2 doping with rare earths by sol–gel method: Photocatalytic activity for phenol degradation. Journal of Sol-Gel Science and Technology. 2010;**56**:219-226. DOI: 10.1007/s10971-010-2297-3

[52] Silveyra R, De La Torre SL, Flores WA, Martínez VC, Elguézabal AA. Doping of TiO_2 with nitrogen to modify the interval of photocatalytic activation towards visible radiation. Catalysis Today. 2005;**107–108**:602-605. DOI: 10.1016/j.cattod.2005.07.023

[53] Yamada K, Yamane H, Matsushima S, Nakamura H, Ohira K, Kouya M, et al. Effect of thermal treatment on photocatalytic activity of N-doped TiO_2 particles under visible light. Thin Solid Films. 2008;**516**:7482-7487. DOI: 10.1016/j.tsf.2008.03.041

[54] Zhang Y, Xu H, Xu Y, Zhang H, Wang Y. The effect of lanthanide on the degradation of RB in nanocrystalline Ln/TiO_2 aqueous solution. Journal of Photochemistry and Photobiology A: Chemistry. 2005;**170**:279-285. DOI: 10.1016/j.jphotochem.2004.09.001

[55] Du P, Bueno-López A, Verbaas M, Almeida AR, Makkee M, Moulijn JA, et al. The effect of surface OH-population on the photocatalytic activity of rare earth-doped P25-TiO_2 in methylene blue degradation. Journal of Catalysis. 2008;**260**:75-80. DOI: 10.1016/j.jcat.2008.09.005

[56] Hwang DW, Lee JS, Li W, Oh SH. Electronic band structure and photocatalytic activity of $Ln_2Ti_2O_7$ (Ln = La, Pr, Nd). he. Journal of Physical Chemistry B. 2003;**107**:4963-4970. DOI: 10.1021/jp034229n

[57] Saif M, Abdel-Mottaleb MSA. Titanium dioxide nanomaterial doped with trivalent lanthanide ions of Tb, Eu and Sm: Preparation, characterization and potential applications. Inorganica Chimica Acta. 2007;**360**:2863-2874. DOI: 10.1016/j.ica.2006.12.052

[58] Zhang Y, Zhang H, Xu Y, Wang Y. Europium doped nanocrystalline titanium dioxide: Preparation, phase transformation and photocatalytic properties. Journal of Materials Chemistry. 2003;**13**:2261-2265. DOI: 10.1039/b305538h

Chapter 8

Assessment of Microbial Load Reduction Efficiency of Sewage Treatment Plants (STP's) in Mysore, Karnataka, India

Severeni Ashili, Harikaranahalli Puttaiah Shivaraju and George Jessen

Abstract

The present study mainly aims to determine the assessment of microbial load reduction efficiency of sewage treatment plants (STPs) in Mysore, Karnataka, India. The raw and treated wastewater samples were collected and tested for irrigation suitability using irrigation indices compared with the Food and Agriculture Organisation (FAO) standards. Seed germination study was carried and the vigour index was reported to be higher for raw sewage although the seedlings treated with this water had wilting shoot tips. The overall results of the present study observed that most of the parameters of both treated and untreated urban wastewaters have exceeded the FAO irrigation standard and continuous usage of such water may cause detrimental effects on the soil and crops. The sewage treatment plants have also shown very low efficiency in microbial load reduction, and this can have health risk implications to the farmers using this effluent.

Keywords: urban wastewater, sewage treatment plant, sewage effluent, vigour index

1. Introduction

Wastewater usage for irrigation is increasingly being practised in developing about 7% of the total irrigated land. Wastewater contains numerous potentially pathogenic microorganisms and a high content of organic matter; therefore, it poses a number of health risks [1–3].

Indirect use of untreated wastewater is one of the most common and extensive types of unintentional wastewater reuse [4–8]; this occurs when untreated wastewater is discharged into freshwater streams, gets diluted and subsequently used by farmers, households and industries. This is a common practice in low- and middle-income countries without or with limited collecting and treatment capacity. There are many pathogenic microorganisms which will always be present in partially treated or untreated wastewater and sewage sludge [9].

This work was primarily aimed to assess the efficiency of the STPs in Mysore in removal of pollutants and microbial pathogens with respect to coliform bacteria to the level set by the FAO, WHO, IBS and CPCB for agricultural and other uses. Secondly, the study tried to evaluate the effect the raw and treated wastewater on the germination and growth of seedlings.

2. Materials and method

2.1 Study area

The study area Mysore city has 887,446 people [10]. Mysore is located at 12° 18′N 76° 39′E 12.30° N 76.65° E and has an average altitude of 770 m (2,526 feet) [11]. The rainy season is from May to October with an average rainfall 782 (697–904) mm. The city has been provided with three wastewater treatment plants (Kesare, Vidarayanapuram and Rayankere). All the treatment plants have facultative aerated lagoons and sedimentation basins [12].

2.2 Sampling and analysis

Assessment of water parameters was carried out in March 2016 to May 2016. Water samples were collected from different stages, i.e., raw wastewater, after facultative pond and final effluent (after sedimentation) in cleaned/sterile 2-L polythene bottles as per the standard methods [13].

These samples were analysed for different physico-chemical, microbial and irrigation quality parameters and agricultural application. Analysis and collection of samples have been done according to standard methods prescribed by the American Public Health Association [13].

2.3 Physico-chemical parameters

Temperature, pH, electrical conductivity (EC), TDS, sodium (Na2+), potassium (K+), calcium (Ca2+), magnesium (Mg2+), total alkalinity (as CaCO3), total hardness (as CaCO3), dissolved oxygen (DO), chemical oxygen demand (COD), biochemical oxygen demand (BOD), etc.

2.4 Microbial parameters

The total coliform count was performed by multiple tube fermentation technique using a set of three tubes inoculated with 10 ml of lactose broth of different strength with samples of 10, 1 and 0.1 ml, respectively [2, 14].

2.5 Irrigation quality parameters

The parameters for irrigation of water quality were calculated based on the result obtained after determination/estimation of sodium (Na2+), potassium (K+), calcium (Ca2+), magnesium (Mg2+) and total alkalinity (as CaCO32−) in mg/l. These values of respected cation and anions were used in the following calculations of the respective parameter of irrigation quality for getting its index or ratios.

2.6 Agricultural application

All the samples were subjected to analysis of various physiochemical parameters with concentration on COD, alkalinity, hardness, K+, Na+, Mg+ and Ca+ followed by the methods of APHA [13].

Three different seeds green gram (*Vigna radiata*), Bengal gram (*Cicer arietinum*) and green peas (*Pisum sativum*) were used for the study. A seedling tray was filled with soil and prepared for seed sowing. The healthy and uniform seeds were cleaned with distilled water, and replicates of five seeds were sown for each of the four different, namely, raw wastewater, final effluent, activated charcoal/sand-filtered effluent and distilled water (control). The seedlings were treated with about 10 ml of sample for twice daily, and also the same was treated with distilled water that was set as control. The sampling study was carried out for a period of 7 days after which the seedlings were studied for different characteristics [15].

Percentage germination was calculated by dividing the number of seedlings germinated with the total number of seedling sown for each treatment set.

$$\%\,germination = \frac{N_G}{N_S} \times 100$$

where N_G is the number of geminated seeds and N_S is the total number of seeds sown.

a. Root length, shoot length, wet weight and dry weight: the root length and shoot length of the germinated seed were measured in centimetre scale. The initial weight of the seedling after 7 days was recorded, and the same was placed in hot air oven for drying at 65°C overnight and weighed. The weight after drying is recorded as dried weight and the weight before drying as the wet weight in grams (g).

b. Seedling vigour index (SVI): vigour index of the seedling was calculated using the formula suggested by Abdul Baki and Anderson in 1973.

$$\text{SVI} = \text{Germination } (\%) \times \text{Seedling Length (Mean Root Length + Shoot Length)}$$

3. Results and discussion

3.1 Physico-chemical parameters

Determination of the general efficiency depends on the overall performances of the different plants in terms of average removal of indicator parameters including TDS, COD, BOD_5, etc. Some of the parameters such as pH directly affect the performance of a secondary treatment process [16] because the existence of most biological life is dependent upon narrow and critical range of pH.

The amount of total dissolved solids (TDS) of inlet as indicated in **Figure 1** ranges from 717 mg/l as observed at Vidarayanapuram STP to 466 mg/l at Kesare STP and has shown little reduction over the stages of all three treatment plants. This proves that aerated lagooning treatment process is inefficient in the removal of TDS. Since the effluent of the three plants is being used for agricultural application or irrigation, the high TDS level may affect the porosity of the soil where this effluent is being used.

The plants have shown an overall biochemical oxygen demand (BOD) removal efficiency of 90, 87 and 97% at Kesare, Vidarayanapuram and Rayankere, respectively. The highest removal is associated with sedimentation tank with an average BOD removal 79% as compared to 54% average removal which was recorded in aerated tank for all plants. Similarly, the plants have an overall average chemical

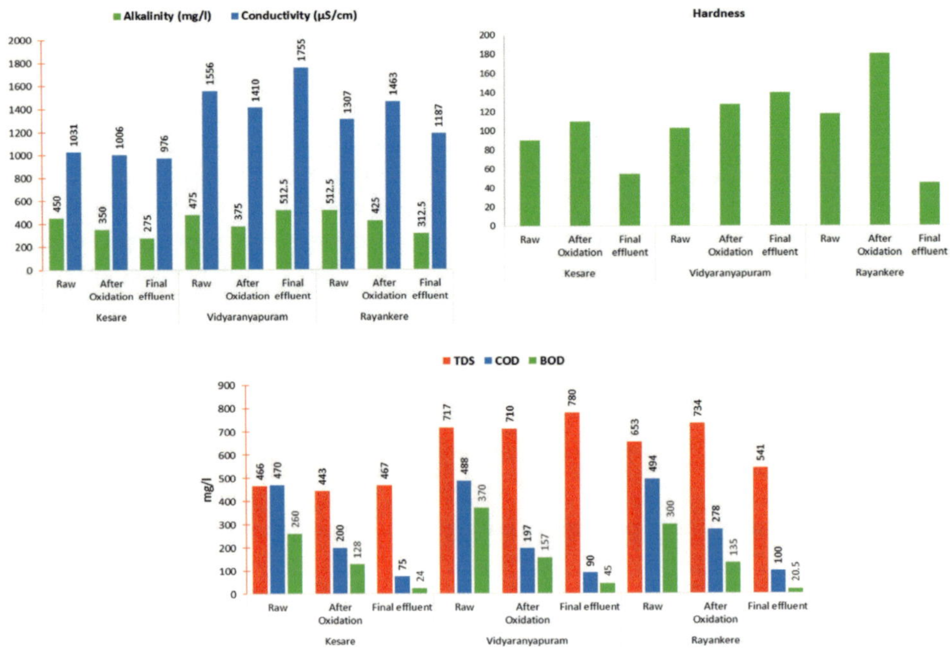

Figure 1.
Physico-chemical stage-wise efficiency of the STPs.

oxygen demand (COD) removal efficiency of 81%, and the high removal is still sedimentation tank with an average removal of 60% as opposed to 53% removal in aeration tank. This makes all the three sewage treatment plants efficient in terms of BOD and COD removal, and they are within the CPCB.

As observed in **Figure 1**, the conductivity of the water has shown no significant reduction over different stages in all the treatment plant. Conductivity increases with the increase of ions, and it is also effectively a surrogate for total dissolved solids and is important for irrigation because it is a measure of the salinity of the water. Salinity is known to restrict water available in the soil for plants to use and also impact crop's physiology and yield.

In the present study, EC of untreated UWW shows a range from 1031 µS/cm at Kesare STP to 1556 µS/cm at Vidyaranyapuram STP. EC of treated UWW showed a range from 976 µS/cm at Kesare STP to 1755 µS/cm at Vidyarayanapuram STP, and the values of untreated and treated UWW at all the three locations lie within the slight to moderate range of the FAO irrigation water quality standards. There was also no significant difference in the alkalinity at different treatment stages of all sewage treatment plants although the water at different location differs significantly in both parameters.

Hardness is usually reported as equivalents of calcium carbonate ($CaCO_3$) and is generally classified as soft, moderately hard, hard and very hard. It is commonly associated with two polyvalent cations, viz. calcium (Ca^{2+}) and magnesium (Mg^{2+}).

As from the results in **Figure 1**, the UWW in Mysore city falls between 75 and 150 mg/l which is classified as moderate hard with the exception of Kesare STP and Rayankere STP final effluent with 55 and 45 mg/l which is in the class of soft water. The effluent from the aerated lagoon in Rayankere STP is also an exception with a total hardness of 180 mg/l which is in the class of hard water as per the US Environmental Protection Agency (EPA). Hardness has no significant effect in the treatment system, but if this water mix with domestic or industrial water sources, it has a significant effect on the amount of detergent and soap required. It also has causes scaling in boiler and other industrial equipment when it is used and therefore reduce equipment overall efficiency.

3.2 Microbial efficiency

The Mysore city UWW system has shown a high number of coliform with Rayankere with the highest CFU of 8.42×10^5 per 100 ml in its raw water (**Figure 2**). All the sewage treatment plants have shown no significant efficiency in the removal of coliform with an overall removal of 5.1, 8.4 and 7.1% for of Kesare STP, Vidyaranyapuram STP and Rayankere STP, respectively. Sedimentation process has proven to be the stage where reduction is high compared to other stages with the highest reduction percentage of 8% at Vidyaranyapuram STP, while other treatment plants have recorded a reduction of less than 5%. This mean that the treatment process used in Mysore city is not effective in removing microbes and therefore the effluent possesses a health risk to farmers.

3.3 Agricultural applicability

3.3.1 Irrigation parameter

Although potassium is not an essential part of any plant component, it is known to be involved and physiological process such as plant water balance and protein synthesis which are important in plant growth. Potassium in UWW originates from human faeces and urine disposal, as human faeces has on average 1.6% and urine has 3.7% (dry weight) potassium. The concentration of potassium of untreated UWW was 18, 19 mg/l for the final effluent and 35 mg/l for the AC/sand column treated effluent as indicated in **Table 1**.

3.3.2 Sodium absorption ratio (SAR)

SAR indicates the effect of relative cation concentration on sodium accumulation in the soil. The present study shows SAR in UWW, final plant effluent and the activated charcoal/sand-filtered effluent to be 11.18, 6.1 and 13.08 respectively (**Table 1**).

SAR in the untreated and AC/sand-filtered water lies in the acute range with respect to sodium irrigation water hazard, while SAR of the treated effluent lies within the slight to moderate hazard. This means that farmers using the reclaimed water need to take extra care whenever they are using this water. Hence, to

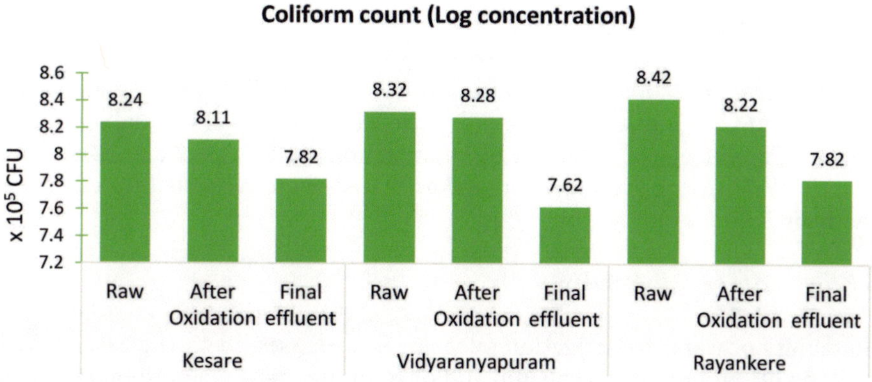

Figure 2.
Coliform count at different treatment stages.

Parameter	Raw	Final Effluent	AC/Sand filtered
Potassium (mg/l)	18	19	35
Sodium (mg/l)	80	51	97
Magnesium (mg/l)	41	56	44
Calcium (mg/l)	61.5	84	66
Carbonates (mg/l)	31.95	142.37	76.4
Bicarbonate (mg/l)	268.28	289.46	155.32
Sodium Absorption Ratio	11.18	6.1	13.08
Residual Sodium carbonate	4.6	3.11	0.54
Soluble Sodium Percentage	48.88	33.49	54.55
Kellys Ratio or Kellys Index	0.78	0.69	0.88

Table 1.
Irrigation parameters of water used in the germination test.

properly assess the suitability of a particular irrigation water supply, the apparent salt tolerance of the specific crop must also be taken into consideration.

3.3.3 Residual sodium carbonate (RSC)

It is another alternative measure of the sodium content in relation with Mg and Ca. Residual sodium carbonate (RSC) exists in irrigation water when the carbonate (CO_3) plus bicarbonate (HCO_3) content exceeds the calcium (Ca) plus magnesium (Mg) content of the water. Where the water RSC is high, extended use of that water for irrigation will lead to an accumulation of sodium (Na) in the soil. If the RSC < 1.25, water is considered safe, while if the RSC > 2.5, the water is not appropriate for irrigation. On this index, both untreated and treated wastewaters are unsafe or inappropriate for irrigation purposes with RSC of 4.6 and 3.11, respectively. The AC/sand-filtered water however lies in the safe range with RSC of 0.56 which is less than 1.25.

3.3.4 Soluble sodium percentage (SSP)

It is also used to evaluate sodium hazard. Water with SSP greater than 60% may result in sodium accumulations that will cause a breakdown of the soil's physical properties. The calculated values of SSP varied from 48.88, 33.49 and 54.55% for untreated UWW, final plant effluent and activated charcoal/sand-filtered effluent, respectively.

3.3.5 Kelly's ratio or Kelly's index

Suitability of water for irrigation purposes is also assessed on the bases of Kelly's ratio. Ratio of sodium versus calcium and of sodium versus magnesium is used as Kelly's ratios. Water having Kelly's ratio of more than one (>1) is considered not suitable for irrigation purposes. Kelly's ratio of the water samples tested was 0.78,

0.69 and 0.88, which implies that this criterion of the groundwater is suitable for irrigation purposes.

3.3.6 Effluent effect on seedling

The germination studies indicate that the percentage germination (**Figure 3**) of each of the three seedlings used for the study varies with different water treatment. The overall high germination rate was observed in raw sewage with 100% germination in both Bengal gram and green gram and 80% germination in green peas, while the least percentage germination was recorded with AC/sand-filtered effluent where a maximum of 80% was recorded for both Bengal gram and green gram and 60% germination in green peas.

This suggest that the percentage germination is not really depended on the concentration of the effluent as it was concluded by the previous study done in Mysore which concluded that the higher concentration of the effluent retards seed germination, whereas the lower concentration may enhance the growth [15]. Bengal gram has the highest germination percentage of 100% almost in all water treatment with only an exception of AC/sand-filtered effluent where only 80% germination was observed. Green peas on the other hand has the lowest overall percentage germination which recorded 80% germination for both control and raw sewage, while only 60% of the seeds germinated in the final effluent and AC/sand-filtered effluent (**Figure 3**).

Seed germination unlike general seedling growth and yield mainly depends on the basic factors such as light, moisture, air, temperature, etc. and therefore is not really affected by pollutant in water. Green gram was observed to have the highest overall growth of 8.65 cm on average under control treatment compared to 6.3 and 8.65 cm of green peas and Bengal gram, respectively; there is however no much difference in the average length of each seedling among treatment (**Figure 3**).

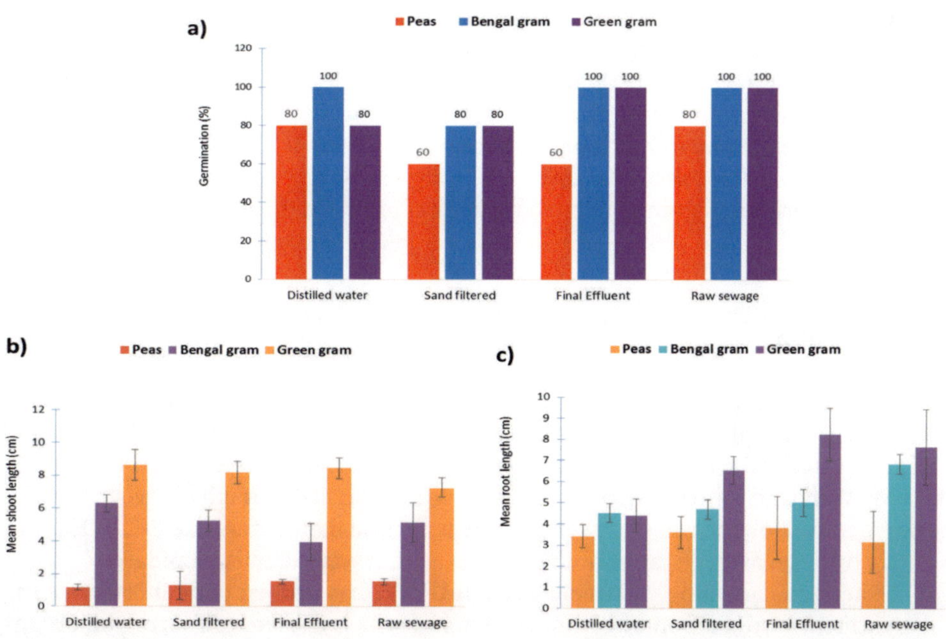

Figure 3.
(a) Percentage of different seeds under different water treatments; (b) mean shoot length of different seeds under different water treatments; and (c) mean root length of different seeds under different water treatments.

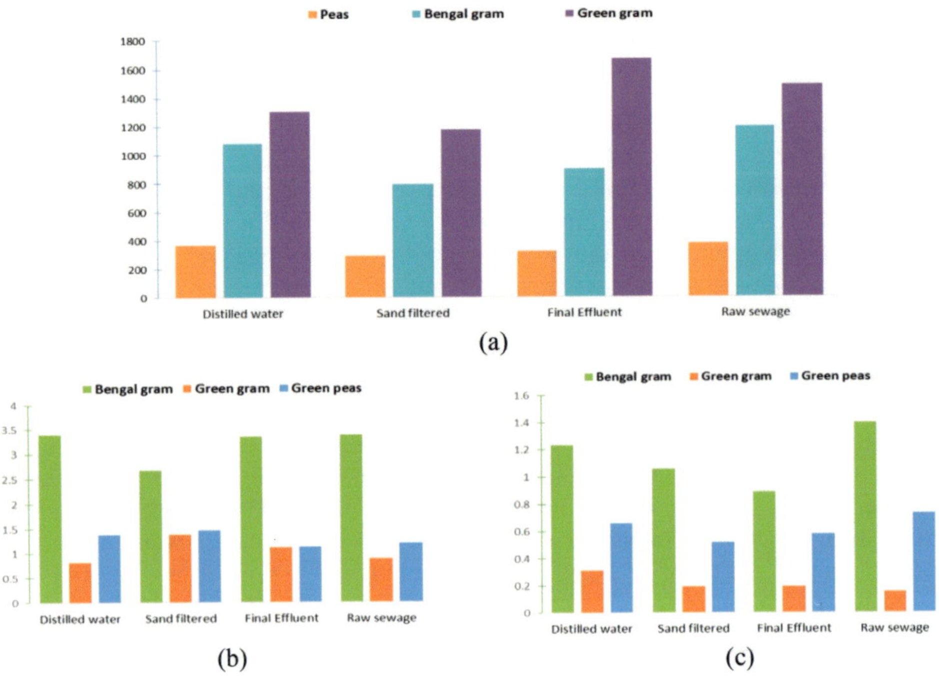

Figure 4.
(a) Vigour index of different seeds under different water treatment; (b) wet weight of different seeds under different water treatments; and (c) dry weight of different seeds under different water treatments.

The treated effluent have the highest average root length for the green peas and green gram seedling of 5.02 and 8.26 cm, respectively, while Bengal gram was high in raw sewage treatment 6.84 cm on average. This high root formation in reclaimed water as opposed to the control can be attributed to the presence of nutrients in it that can act as plant growth enhancer. Overall, green gram has shown a highest vigour, while green peas have the lowest vigour across all treatment. It is also important to note that among all the treatment the highest were recorded in the final effluent and raw sewage, and this can still be attributed to the nutrient and ion content that is associated with this water (**Figure 4(a)**).

Both wet weight and dry weight as indicated in **Figure 4(b)** and **(c)** have no relation with water treatment parameter. The highest weight recorded in both wet and dry was for Bengal gram, and green gram has recorded the lowest wet and dry weight.

4. Conclusion

The aerated lagooning treatment system used has proven to be effective in the reduction of some of the parameters and ineffective in some. Parameters like BOD and COD of treated effluent in UWW are within the CPCB permissible limit for disposing the UWW on land for irrigation. Continuous use of this effluent may however have a negative impact on the soil since some of the irrigation parameters like EC, TDS and SAR are over the average range of the FAO irrigation water quality standards.

The aerated lagooning treatment used in UWW has proven to be ineffective in the treatment or reduction of coliform bacterial which is used as a surrogate organism for microbial pathogens. This may be partially due to the chosen treatment system, but the operation and maintenance can also have an effect on the efficiency of the treatment.

Author details

Severeni Ashili[1,2]*, Harikaranahalli Puttaiah Shivaraju[2] and George Jessen[2,3]

1 Department of Mathematics, Science and Sports Education, University of Namibia, Namibia

2 Department of Water and Health, Faculty of Life Sciences, JSS University, Mysore, Karnataka, India

3 Eben-Ezer Degree College, Eben-Ezer Group of Institutions, Bangalore, Karnataka, India

*Address all correspondence to: severeni.ashili@gmail.com

References

[1] Howard I, Espigares E, Lardelli P, Martín JL, Espigares M. Evaluation of microbiological and physico-chemical indicators for wastewater treatment. Environmental Toxicology. 2004;**19**(3): 241-249. Available from: https://onlinelibrary.wiley.com/doi/abs/10.1002/tox.20016

[2] George J, Divya L, Magesh SB, Suriyanarayanan S. An assessment of removal efficiency for the bacterial pathogens in Mysore urban water treatment system, Karnataka, India: A case study. Desalination & Water Treatment. 2016;**57**(23):10886-10893. DOI: 10.1080/19443994.2015.1039601

[3] George J, Lakshminarayanan D, Ashili S, Sarvajayakesavalu S. Urban wastewater treatment systems: Assessment of removal efficiency based on microbial pathogens: A case study in Mysore, India. In: Innovations in Agricultural and Biological Engineering, Sustainable Biological Systems for Agriculture, Emerging Issues in Nanotechnology, Biofertilizers, Wastewater, and Farm Machines. USA: Apple Academic Press; 2018. pp. 269-279. Available from: https://www.taylorfrancis.com/books/e/9781351676595/chapters/10.1201%2F9781315165264-13

[4] Jimenez B, Asano T. Water reclamation and reuse around the world. In: Jimenez B, Asano T, editors. Water Reuse: An International Survey of Current Practice, Issues and Needs. London: IWA; 2008. pp. 1-26. Available from: http://www.scirp.org/(S(351jmbntvnsjt1aadkposzje))/reference/ReferencesPapers.aspx?ReferenceID=1294610

[5] Keraita B, Jimenez B, Drechsel P. Extent and implications of agricultural reuse of untreated, partly treated, and diluted wastewater in developing countries. CAB Reviews: Perspectives in Agriculture, veterinary Science, Nutrition and Natural Resources. 2008;**3**(58):1-5. Available from: http://www.ovid.com/site/catalog/databases/3292.jsp

[6] Suriyanarayanan S, George J, Divya L, Balasubramanian S. Effect of waste paper based paper industry effluents on the growth of tree seedlings. Journal of Environmental Research and Development. 2012;**7**:1117-1126. Available from: http://www.jerad.org/ppapers/dnload.php?vl=7&is=2A&st=1117

[7] Midhun G, Divya L, George J, Jayakumar P, Suriyanarayanan S. Wastewater treatment studies on free water surface constructed wetland system. In: Prashanthi M, Sundaram R, editors. Integrated Waste Management in India: Status and Future Prospects for Environmental Sustainability. Switzerland: Springer International Publishing; 2016. pp. 97-109. ISBN: 978-3-319-27226-9

[8] George J, Divya L, Suriyanarayanan S. Quantitative microbial risk assessment in the management of *Escherichia coli* strains via drinking water. Journal of Environmental Research and Development. 2013;**8**:60-68. Available from: http://www.jerad.org/ppapers/dnload.php?vl=8&is=1&st=60

[9] Walls K. Health implications of increasing reuse of wastewater as an adaption to climate change. Journal of Environmental Engineering and Ecological Science. 2015;**4**:2. DOI: 10.7243/2050-1323-4-2

[10] Ministry of Home Affairs. Census India. Office of the Registrar General & Census Commissioner, Government of India. 2011. Available from: http://www.censusindia.gov.in [Accessed: 25-10-2018]

[11] George J, An W, Joshi D, Zhang D, Yang M, Suriyanarayanan S. Quantitative microbial risk assessment to estimate health risk in urban drinking water systems of Mysore, Karnataka, India. Exposure & Health. 2015;**7**(3): 331-338. Available from: https://link.springer.com/article/10.1007/s12403-014-0152-4

[12] Divya L, Jessen G, Suriyanarayanan S, Karthikeyan K. Studies on pathogenic bacterial strains from selected Sewage Treatment Plants (STPs) of Mysore, Karnataka, India during different seasons: A comparative appraisal. Journal of Environmental Research and Development. 2014;**9**(1):24-30. Available from: http://www.jerad.org/ppapers/dnload.php?vl=9&is=1&st=24

[13] APHA, AWWA and WEF. Standard Methods for the Examination of Water and Wastewater. 21st ed. Washington: American Public; 2005

[14] APHA. Standard Methods for the Examination of Water and Wastewater. 20th ed. Washington DC: American Public Health Association Inc; 1998

[15] Divya L, George J, Midhun G, Magesh SB, Suriyanarayanan S. Impacts of treated sewage effluent on seed germination and vigour index of monocots and dicot seeds. Russian Agricultural Sciences. 2015;**41**(4):252-257. Available from: https://link.springer.com/article/10.3103/S1068367415040242

[16] Metcalf & Eddy. Wastewater Engineering: Treatment and Resource Recovery. 5th ed. New York: McGraw-Hill; 2014